# METHODS IN MOLECULAR BIOLOGY

*Series Editor*
John M. Walker
School of Life and Medical Sciences
University of Hertfordshire
Hatfield, Hertfordshire, AL10 9AB, UK

For further volumes:
http://www.springer.com/series/7651

# Kidney Organogenesis

## Methods and Protocols

Edited by

**Seppo Vainio**

*Biocenter Oulu, Laboratory of Developmental Biology, Oulu University, Oulu, Finland;*
*InfoTech Oulu, Oulu University, Oulu, Finland;*
*Borealis Biobank of Northern Finland, Oulu University, Oulu, Finland;*
*Faculty of Biochemistry and Molecular Medicine, Oulu University, Oulu, Finland*

Humana Press

*Editor*
Seppo Vainio
Biocenter Oulu, Laboratory of Developmental Biology
Oulu University
Oulu, Finland

InfoTech Oulu
Oulu University
Oulu, Finland

Borealis Biobank of Northern Finland
Oulu University
Oulu, Finland

Faculty of Biochemistry and Molecular Medicine
Oulu University
Oulu, Finland

ISSN 1064-3745 ISSN 1940-6029 (electronic)
Methods in Molecular Biology
ISBN 978-1-4939-9020-7 ISBN 978-1-4939-9021-4 (eBook)
https://doi.org/10.1007/978-1-4939-9021-4

Library of Congress Control Number: 2018966342



Cover Caption: Optical projection tomography-based visualization of the nephrons (in red) and the collecting duct tree (in grey). Automated spot detection function of the Imaris program was applied to highlight the nephrons (in yellow). E16.5. mouse kidney.

This Humana Press imprint is published by the registered company Springer Science+Business Media, LLC, part of Springer Nature.
The registered company address is: 233 Spring Street, New York, NY 10013, U.S.A.

# Preface

The mammalian kidney has served as a model to study the fundamental developmental mechanisms from the late 1950s when Clifford Grobstein established the capacity to culture the microdissected kidney rudiments. By using such microsurgery approaches and organ culture technologies, Lauri Saxén's team showed later that like in the primary embryonic induction, the process of organogenesis involves signals that mediate the secondary inductive tissue interactions. In the context of the kidney, signaling occurs between the ureteric bud and the metanephric mesenchyme. These coordinate the stepwise kidney morphogenesis and the associated cell differentiation.

Even though the powers to use the kidney as a developmental model system were established more than 60 years ago, the classic ex vivo kidney culture approach was technologically still very limited. Especially no ways existed to assay the roles of the molecular players involved. Pioneers such as Peter Ekblom used initially specific antibodies and small molecule inhibitors to target the molecular mediators, but the actual gene functions could not be approached.

When the embryonic stem cell culture and the gene targeting technologies became a routine, this opened the era of the renal functional genetics. By using the gene knockout approach, many genes were identified to be essential for kidney ontogenesis. Evidence could be finally reached now that the signals that mediate the kidney tubule induction are encoded by the *Wnt* gene family. The Wnts serve as the key mediators in the classic tubule induction model system and in vivo signaling from the ureteric bud to the metanephric mesenchyme to trigger nephrogenesis. The renal tissue-specific gene knockout capacity was a step forward, and to date gene functions can be inactivated from the major kidney tissue compartments. At this stage, the ex vivo kidney organ culture models provided again a valuable tool to depict the epistasis relationships of the genes in organogenesis.

During the recent years, the kidney tissue engineering methods to study the mechanisms in tissue biogenesis have undergone a major revolution. The mammalian kidney rudiment can be introduced to single cells. However, kidney developmental capacity can be reconstituted thereafter. The kidney progenitor cells can also be frozen, and the cell fates can be studied at the single-cell level as well.

The 4D imaging technologies together with the availability of wealth of the gene-engineered renal indicator developmental and disease models have opened a series of experimental studies. The molecular genetic technologies have developed to a sophistication that the involved genetic programs can be studied also at the single-cell level. Importantly, functions of specific genes and candidate renal developmental factors can be approached in the ex vivo settings as well.

Major advancements have taken place also in the renal stem cell technologies. Protocols have become available to program the pluripotent cells to the defined cell lineages such as the ones of the kidney. Such protocols are in our toolbox for the mouse and human. The skills to take a good use of the patient-derived cells with a kidney disease are a historical opening in human kidney-related medical research. Such patient-derived cells can also be stored in the biobanks to gain more value. Importantly kidney organoids can be generated from such engineered cells, and the CRISPR gene editing can be achieved in these cells as well. These approaches can now be combined to establish large-scale gene editing microfluidics and robotic-automated platforms.

The current Springer book contains a series of protocols that allow their user to get access both to the classic and the very modern ex vivo mammalian developmental kidney model systems. The book contains chapters that illustrate the elegant 3D and 4D imaging modalities, the powers of the kidney decellularizing, and how to build the kidney tissues from the reprogrammed cells.

The pioneers of the field are likely thrilled of the capacities that the kidney community has today in their use. At their time they could only dream of them. Via the combination of the classic and the modern renal technologies, we will be now able to generate more efficiently value for the patients that look for new therapies for their kidney symptoms.

***Oulu, Finland*** ***Seppo Vainio***

# Contents

## Contributors

TOSHIKAZU ARAOKA • *Center for iPS Cell Research and Application (CiRA), Kyoto University, Kyoto, Japan*
JUAN CARLOS IZPISUA BELMONTE • *Gene Expression Laboratory, The Salk Institute for Biological Studies, La Jolla, CA, USA*
AARON C. BROWN • *Center for Molecular Medicine, Maine Medical Center Research Institute, Scarborough, ME, USA*
PABLO CABRERO • *Institute of Molecular, Cell and Systems Biology, College of Medical, Veterinary and Life Sciences, University of Glasgow, Glasgow, Scotland, UK*
SANDHANAKRISHNAN CATTAVARAYANE • *Faculty of Veterinary Medicine, Section of Pathology, Department of Veterinary Biosciences, University of Helsinki, Helsinki, Finland*
XAVIER CAUBIT • *Aix Marseille University, CNRS, IBDM, Marseille, France*
THERESA CHOW • *Lunenfeld-Tanenbaum Research Institute, Mt. Sinai Hospital, Toronto, ON, Canada; Department of Physiology, University of Toronto, Toronto, ON, Canada*
GUILLERMO MARTINEZ CORRALES • *Institute of Molecular, Cell and Systems Biology, College of Medical, Veterinary and Life Sciences, University of Glasgow, Glasgow, Scotland, UK*
FRANK COSTANTINI • *Department of Genetics and Development, 1418 Hammer Health Sciences Center, Columbia University, New York, NY, USA*
JAMIE A. DAVIES • *Deanery of Biomedical Sciences, University of Edinburgh, Edinburgh, UK*
SHIREEN-A. DAVIES • *Institute of Molecular, Cell and Systems Biology, College of Medical, Veterinary and Life Sciences, University of Glasgow, Glasgow, Scotland, UK*
ANTHONY J. DORNAN • *Institute of Molecular, Cell and Systems Biology, College of Medical, Veterinary and Life Sciences, University of Glasgow, Glasgow, Scotland, UK*
JULIAN A. T. DOW • *Institute of Molecular, Cell and Systems Biology, College of Medical, Veterinary and Life Sciences, University of Glasgow, Glasgow, Scotland, UK*
MONA ELHENDAWI • *Deanery of Biomedical Sciences, University of Edinburgh, Edinburgh, UK; Faculty of Medicine, Clinical Pathology Department, Mansoura University, El-Mansoura, Egypt*
LAURENT FASANO • *Aix Marseille University, CNRS, IBDM, Marseille, France*
SAURAV GHIMIRE • *Institute of Molecular, Cell and Systems Biology, College of Medical, Veterinary and Life Sciences, University of Glasgow, Glasgow, Scotland, UK*
ASHWANI K. GUPTA • *Center for Molecular Medicine, Maine Medical Center Research Institute, Scarborough, ME, USA*
KRITHIKA HARIHARAN • *Berlin–Brandenburg Center for Regenerative Therapies (BCRT), Charité - Universitätsmedizin Berlin, Berlin, Germany*
ANNELIIS IHERMANN-HELLA • *GM-unit at Helsinki Institute of Life Science and Medicum, University of Helsinki, Helsinki, Finland*
ANDREAS KURTZ • *Berlin–Brandenburg Center for Regenerative Therapies (BCRT), Charité - Universitätsmedizin Berlin, Berlin, Germany*
SATU KUURE • *GM-unit at Helsinki Institute of Life Science and Medicum, Institute of Biotechnology, University of Helsinki, Helsinki, Finland*
MELANIE L. LAWRENCE • *Deanery of Biomedical Sciences, University of Edinburgh, Edinburgh, UK*

ZHONGWEI LI • *Division of Nephrology and Hypertension, Department of Medicine and USC/UKRO Kidney Research Center, Keck School of Medicine of the University of Southern California, Los Angeles, CA, USA; Department of Stem Cell Biology and Regenerative Medicine, Keck School of Medicine of the University of Southern California, Los Angeles, CA, USA*
FILIPA M. LOPES • *Faculty of Biology Medicine and Health, Division of Cell Matrix Biology and Regenerative Medicine, School of Biological Sciences, University of Manchester, Manchester, UK*
SHIN-ICHI MAE • *Center for iPS Cell Research and Application (CiRA), Kyoto University, Kyoto, Japan*
AKI MANNINEN • *Faculty of Biochemistry and Molecular Medicine, Biocenter Oulu, Oulu Center for Cell-Matrix Research, University of Oulu, Oulu, Finland*
RICHARD MARLEY • *Institute of Molecular, Cell and Systems Biology, College of Medical, Veterinary and Life Sciences, University of Glasgow, Glasgow, Scotland, UK*
RYUICHI NISHINAKAMURA • *Department of Kidney Development, Institute of Molecular Embryology and Genetics, Kumamoto University, Kumamoto, Japan*
KENJI OSAFUNE • *Center for iPS Cell Research and Application (CiRA), Kyoto University, Kyoto, Japan*
LEIF OXBURGH • *Center for Molecular Medicine, Maine Medical Center Research Institute, Scarborough, ME, USA*
ANDREW S. POTTER • *Division of Developmental Biology, Cincinnati Children's Medical Center, Cincinnati, OH, USA*
S. STEVEN POTTER • *Division of Developmental Biology, Cincinnati Children's Medical Center, Cincinnati, OH, USA*
RENATA PRUNSKAITE-HYYRYLÄINEN • *Faculty of Biochemistry and Molecular Medicine, University of Oulu, Oulu, Finland*
ALEKSANDRA RAK-RASZEWSKA • *Faculty of Biochemistry and Molecular Medicine, University of Oulu, Oulu, Finland*
PETRA REINKE • *Berlin–Brandenburg Center for Regenerative Therapies (BCRT), Charité - Universitätsmedizin Berlin, Berlin, Germany; Department of Nephrology and Intensive Care, Charité - University Medicine Berlin, Berlin, Germany*
IAN M. ROGERS • *Lunenfeld-Tanenbaum Research Institute, Mt. Sinai Hospital, Toronto, ON, Canada; Department of Physiology, University of Toronto, Toronto, ON, Canada; Department of Obstetrics and Gynecology, University of Toronto, Toronto, ON, Canada*
MAKOTO RYOSAKA • *Center for iPS Cell Research and Application (CiRA), Kyoto University, Kyoto, Japan*
ULLA SAARELA • *Faculty of Biochemistry and Molecular Medicine, University of Oulu, Oulu, Finland; Laboratory of Developmental Biology, Biocenter Oulu, InfoTech, Oulu, Finland; Oulu Center for Cell Matrix Research, Oulu, Finland*
IRENE SANCHEZ-MARTIN • *Aix Marseille University, CNRS, IBDM, Marseille, France*
FRANZ SCHAEFER • *Department of Medicine, University of Heidelberg, Heidelberg, Germany*
ABHISHEK SHARMA • *Faculty of Biochemistry and Molecular Medicine, Biocenter Oulu, University of Oulu, Oulu, Finland*
ILYA SKOVORODKIN • *Faculty of Biochemistry and Molecular Medicine, University of Oulu, Oulu, Finland; Laboratory of Developmental Biology, Biocenter Oulu, InfoTech, Oulu, Finland; Oulu Center for Cell Matrix Research, Oulu, Finland*
ATSUHIRO TAGUCHI • *Department of Genome Regulation, Max Planck Institute for Molecular Genetics, Berlin, Germany*

PAULI TIKKA • *Department of Medicine, University of Heidelberg, Heidelberg, Germany*
SEPPO VAINIO • *Biocenter Oulu, Laboratory of Developmental Biology, Oulu University, Oulu, Finland; InfoTech Oulu, Oulu University, Oulu, Finland; Borealis Biobank of Northern Finland, Oulu University, Oulu, Finland; Faculty of Biochemistry and Molecular Medicine, Oulu University, Oulu, Finland*
JENNIFER WHITELEY • *Lunenfeld-Tanenbaum Research Institute, Mt. Sinai Hospital, Toronto, ON, Canada*
ADRIAN S. WOOLF • *Faculty of Biology Medicine and Health, Division of Cell Matrix Biology and Regenerative Medicine, School of Biological Sciences, University of Manchester, Manchester, UK; Royal Manchester Children's Hospital, Manchester University NHS Foundation Trust, Manchester Academic Health Science Centre, Manchester, UK*
YASUHIRO YOSHIMURA • *Department of Kidney Development, Institute of Molecular Embryology and Genetics, Kumamoto University, Kumamoto, Japan*
KAI ZHANG • *Faculty of Biochemistry and Molecular Medicine, Biocenter Oulu, Oulu Center for Cell-Matrix Research, University of Oulu, Oulu, Finland*

# Part I

## Generation of Mosaic Embryonic Kidneys and Culture of the Kidney Tissues Ex Vivo

# Chapter 1

# Generating Genetic Mosaic Mouse Embryos or Organoids for Studies of Kidney Development

## Frank Costantini

## Abstract

For studies of gene function during development, it can be very useful to generate mosaic embryos in which a small subset of cells in a given cell lineage lacks a gene of interest and carries a marker that allows the mutant cells to be specifically visualized and compared to wild-type cells. Several methods have been used to generate genetically mosaic mouse kidneys for such studies. These include (1) chimeric embryos generated using embryonic stem cells, (2) chimeric renal organoids generated by dissociation and reaggregation of the fetal kidneys, (3) generation of a knockout allele with a built-in reporter gene, (4) mosaic analysis with double markers (MADM), and (5) mosaic mutant analysis with spatial and temporal control of recombination (MASTR). In this chapter, these five methods are described, and their advantages and disadvantages are discussed.

**Key words** Mosaic, Chimera, MADM, MASTR, Knockout, Conditional knockout, Organoid, Cre recombinase, FLP recombinase

## 1 Introduction

Knockout (KO) mice have been invaluable for studies of the genes that are important for kidney development, as for all areas of developmental biology. However, the KO of some genes involved in kidney development results in a failure of urogenital system development before the adult (metanephric) kidney forms or very early in the process. For example, KO of *Lhx1* results in a defect in nephric duct (ND) development, so the ureteric bud (UB) rarely has a chance to form [1], while KO of *Ret* causes an early failure in UB formation or branching [2]. In such cases, it can be difficult to discern the specific roles in kidney development of the gene in question from the mutant phenotype. In some cases, a solution may be to use conditional knockout (CKO) mice [3], in which the gene of interest is flanked by loxP sites ("floxed") and deleted in a specific sub-lineage of kidney cells using Cre recombinase, driven by a lineage-specific gene. In the case of *Lhx1*, for example, deletion

Seppo Vainio (ed.), *Kidney Organogenesis: Methods and Protocols*, Methods in Molecular Biology, vol. 1926,
https://doi.org/10.1007/978-1-4939-9021-4_1, © Springer Science+Business Media, LLC, part of Springer Nature 2019

of a CKO allele, using a Cre recombinase gene expressed only in the nephron cell lineage, allowed the early defect in ND development to be bypassed, revealing another role of *Lhx1* in nephrogenesis [4]. However, even lineage-specific CKO of certain genes can have major consequences that lead to an early failure of organogenesis.

For these reasons, it is often informative to generate genetic mosaic kidneys, in which only a small proportion of cells have deleted the gene of interest; in such cases, it is also important that the mutant cells be differentially labeled, so their behaviors and fates can be followed during organogenesis and distinguished from their wild-type (WT) neighbors. Genetic mosaics have been used for many decades in mouse developmental biology [5] and even more extensively in lower organisms such as flies and worms. Here, we review five methods for generating mosaics that have been useful for studies of kidney development. One limitation to keep in mind (which applies to all methods of generating mosaics) is that these methods are useful primarily for genes with a cell-autonomous function, such as those encoding transcription factors, receptors, or intracellular signaling molecules, and less so for cell-non-autonomous genes, such as those encoding secreted proteins. This is because isolated cells that are mutant for a secreted protein may be "rescued" by secretion of the protein by their neighboring cells and thus may not display a mutant phenotype.

## 2 Chimeric Embryos Generated Using Embryonic Stem Cells

Chimeric mouse embryos (consisting of cells derived from more than one embryo, with different genotypes) can be generated by several methods. However, the most useful method for this purpose is to (1) derive embryonic stem (ES) cells that are homozygous for a KO gene of interest and then (2) microinject the mutant ES cells into WT host blastocysts (or alternatively, aggregate the ES cells with morula-stage host embryos) [6] (Fig. 1). The ES cells should also carry a "reporter gene," expressing a fluorescent protein such as GFP (or another marker protein such as beta-galactosidase) in the cell type(s) to be studied. If possible, the host embryos should carry a different reporter allele. For example, in studies of the role of the *Ret* gene in UB branching [7, 8], the ES cells were engineered to be homozygous for a *Ret*-KO allele and also to carry the transgene *Hoxb7/GFP*, which is expressed specifically in the ND and UB cell lineage; and the host WT embryos were engineered to express cyan fluorescent protein (CFP) in the same cell lineage (*see* **Note 1**). This allowed the *Ret*-mutant (GFP+) and the wild-type (CFP+) cells in the developing ND and UB to be visualized and distinguished. These studies revealed that *Ret* is important for ND cell rearrangements that allow a subset of ND

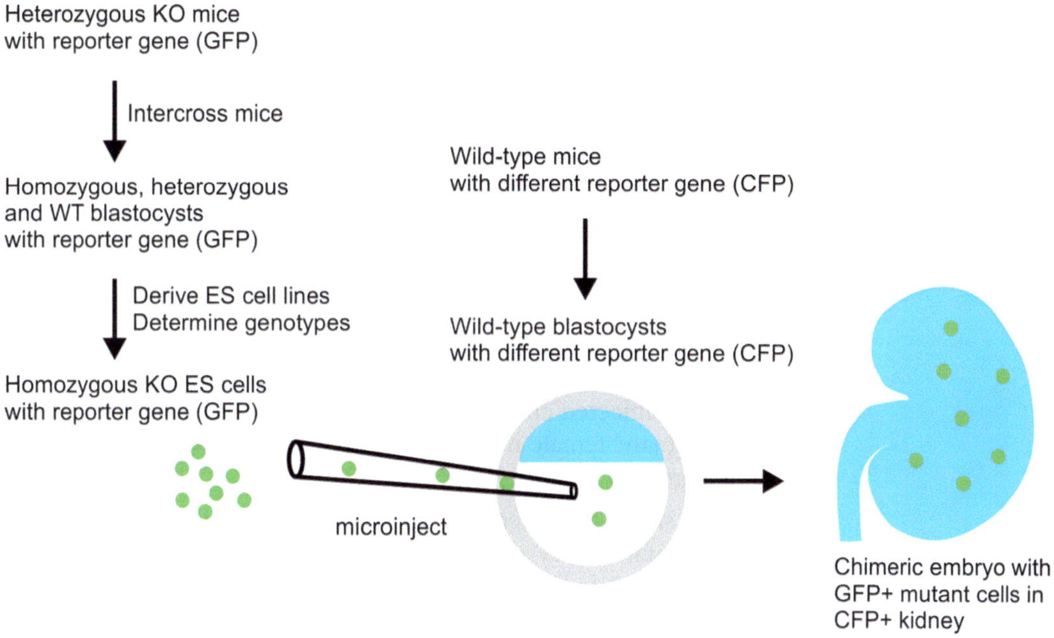


**Fig. 1** Generation of chimeric mice with genetically mosaic kidneys, via derivation of mutant ES cells. The mutant ES cells carry a reporter gene (e.g., GFP) and are microinjected into wild-type host blastocysts with a different reporter gene (e.g., CFP). The reporter genes can be uniformly expressed as shown here, or they can be lineage-specific. See text for details

cells to generate the "tip domain" of the primary UB as it emerges from the ND [8].

### *2.1 Materials and Methods*

1. Breed a colony of mice carrying the KO gene to be studied, as well as a reporter gene encoding a fluorescent protein, or other marker, that is expressed in the cell lineage or cell type to be studied. If the KO gene is recessive-lethal, the animals will have to be heterozygous.
2. Obtain blastocyst-stage embryos by intercrossing the appropriate mice, so that at least a fraction of the embryos will be homozygous for the KO gene and will also carry the reporter gene.
3. Generate an ES cell line from each individual embryo [6, 9], using enough embryos so that at least one ES cell line (preferably two or more—*see* **Note 2**) will be homozygous for the KO gene (*see* **Note 3**) and will also carry the reporter gene. For example, if the embryos are generated by intercrossing KO heterozygotes, and one parent is also heterozygous for the reporter gene, ~1/8 embryos (and therefore ~1/8 ES cell lines) will have the desired homozygous KO genotype and also carry the reporter. In such a case, it is advisable to generate about 15–20 ES cell lines, so it is likely to obtain two or more with the right genotype. Also *see* **Note 4** concerning the potential use of CRISPR/Cas9 [10] to generate the mutant ES cells.

4. It is important to also generate ES cells that are WT for the gene being studied, and that carry the same reporter gene, to generate *control* chimeric embryos.
5. Determine the genotypes of the ES cell lines, to identify at least one and preferably two with the desired mutant genotype. Cryopreserve the ES cells for later recovery.
6. Generate host embryos that are WT for the gene being studied (at blastocyst stage for ES cell microinjection, or morula stage for ES cell aggregation).
7. Microinject the ES cells into the host blastocysts (or aggregate the ES cells into host morulae), and surgically re-implant the microinjected embryos into the reproductive tracts of pseudo-pregnant female mice so that they can develop to the desired stage [6].
8. To study the effects of the KO on the mutant cells during prenatal kidney development, recover the chimeric embryos at the desired stage of gestation and examine the properties of the mutant cells (e.g., number, location, or other characteristics) in the kidney, in comparison with neighboring host embryo cells, and with control ES-derived cells in control chimeras. An advantage of using fluorescent reporter proteins, rather than beta-galactosidase or other histochemical markers, is that the mutant cells can be followed at multiple time points by time-lapse imaging of live kidney cultures [11].

### 2.2 Advantages

1. Virtually any gene for which KO mice exist can be studied in this way, and novel KO ES cell lines can easily be generated using the CRISPR/Cas9 system (*see* **Note 4**).
2. Once the appropriate ES cell lines are generated, an unlimited number of chimeric embryos can be generated.

### 2.3 Disadvantages

1. Deriving ES cells and generating chimeric embryos are technically difficult, require specialized equipment and expertise, and are expensive.
2. The percentage of mutant cells in the chimeras is difficult to control (*see* **Note 5**) and is often higher than desirable (usually 20–50%).

## 3 Chimeric Renal Organoids Generated by Dissociation and Reaggregation of Fetal Kidneys

If fetal kidneys (typically E11.5–E13.5) are dissociated to single cells, and the cells are allowed to reaggregate and grow in culture, they re-form kidney-like organoids that develop branched ureteric bud tubules and nephrons [12–15]. By mixing kidney cells from

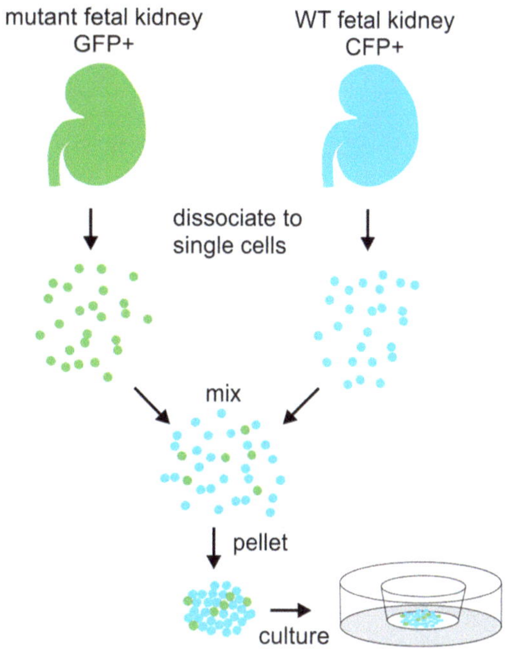


**Fig. 2** Generation of chimeric renal organoids by dissociation and reaggregation of fetal kidneys. The diagram illustrates an experiment with kidneys uniformly expressing GFP or CFP, but fluorescent protein genes expressed in specific cell lineages can also be used. If one of the kidneys is homozygous for a mutant gene, this method can be used to study how that gene influences cell fate during renal organoid development

embryos of two different genotypes, with appropriate marker genes to distinguish them, genetically mosaic kidney organoids can be generated (Fig. 2). This approach has been used to study the effects of mutations in *Spry1* and *Fgfr2* on UB branching morphogenesis [16].

### 3.1 Materials and Methods

(Since the detailed methods are published elsewhere [13, 16, 17], this is only an outline of the steps involved.)

1. Generate, by breeding, mouse embryos that are homozygous for the KO gene of interest and other embryos that are WT, each carrying a different reporter gene (e.g., GFP, CFP, RFP, etc.) that is expressed in the cell lineage of interest (e.g., UB, nephron, stroma). *See* **Note 6**.
2. Dissect the kidneys and dissociate them to single cells.
3. Count the cells, and mix cells of the two genotypes in the desired proportions.
4. Reaggregate the cells by centrifugation, and place the cell pellet on a Transwell filter.

5. Culture the mosaic organoids, and photograph at the desired times to monitor the behaviors of the mutant vs. the wild-type cells.

### 3.2 Advantages

1. Virtually any gene for which KO mice exist can be studied in this way.
2. No specialized equipment (e.g., for microinjection) or embryo transfer surgery is required.
3. The starting percentage of mutant vs. WT cells can be precisely controlled.

### 3.3 Disadvantages

1. If the KO causes a severe defect early in kidney development, it may not be feasible to obtain sufficient numbers of mutant kidney cells to perform the experiments.
2. The development of renal organoids (even using WT kidney cells) is not completely normal (e.g., multiple UB trees are usually generated) so this can only be used to study processes that occur in such cultured organoids. However, this can serve as a first approach before using another more expensive, time-consuming, or difficult method of generating mosaics.

## 4 Generation of a Knockout Allele with a Built-in Reporter Gene

It is easy to generate mosaic kidneys using a CKO allele together with a tamoxifen-inducible Cre gene, by giving a low dose of tamoxifen. However, in most cases such mosaic organs are of limited experimental value because the mutant cells cannot be easily distinguished from the wild-type cells. One solution is to custom-design and generate mice with a CKO allele that carries a built-in reporter gene (also known as a "conditional reporter"). For example, Uesaka et al. [18] generated a CKO allele of *Gfra1* (a gene encoding a co-receptor for GDNF that functions in concert with the Ret receptor tyrosine kinase) in which, upon Cre-mediated recombination, *Gfra1* expression is lost and, simultaneously, a previously inactive *GFP* gene is placed under the control of the *Gfra1* promoter (Fig. 3). In this way, only the cells which have deleted *Gfra1* will express GFP. These mice were used to study the importance of *Gfra1* for UB branching during kidney development [19].

### 4.1 Materials and Methods

1. Design and construct the targeting vector for the gene of interest. Such an allele could be constructed in several ways, and the design used by Uesaka et al. [18] is described in Fig. 3.
2. Use the targeting vector to generate targeted ES cells.

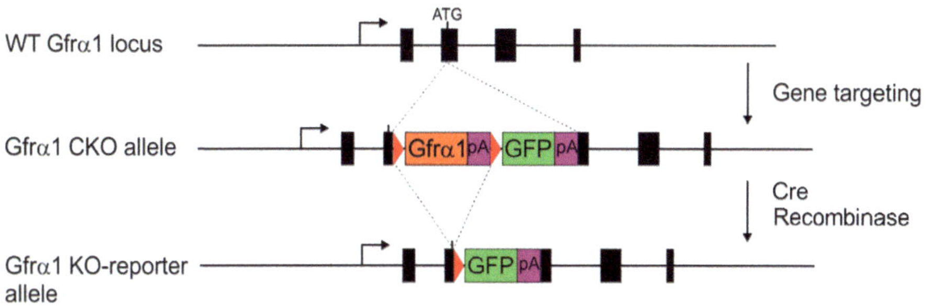


**Fig. 3** Generation of a *Gfra1* conditional knockout allele with a built-in reporter gene (or "conditional reporter"). Gene targeting in ES cells was used to introduce the following sequences into exon 2 of the *Gfra1* gene, following the initiation codon: loxP site (red triangle), Gfrα1 cDNA, polyadenylation site (pA), loxP site, GFP cDNA, and polyadenylation site. In this CKO allele, while the endogenous *Gfra1* gene is disrupted, Gfrα1 is still expressed from the cDNA sequence, so the allele is functional, while GFP expression is blocked by the preceding pA site. When Cre catalyzes recombination between the two loxP sites, the Gfrα1 sequence and first pA sequence are deleted, and the GFP gene is now expressed from the *Gfra1* promoter. Diagram modified from Uesaka et al. [18] where more details can be found

3. Use the targeted ES cell to generate germ-line chimeric mice, and then breed to obtain a strain of mice carrying the new allele.
4. Cross the mice with an appropriate CreER (tamoxifen-inducible Cre) [20] line, and induce with a low dose of tamoxifen to induce recombination in a small proportion of cells. The dose will vary for different alleles and different CreER lines and must be determined empirically.

### 4.2 Advantages

1. Once the mouse strain has been generated, it is easy to generate mosaic embryos.
2. The degree of mosaicism in the cell type(s) of interest can be adjusted by the choice of a CreER gene and by varying the dose of tamoxifen. While high doses of tamoxifen can cause embryonic lethality, usually low doses are sufficient to generate mosaic recombination.

### 4.3 Disadvantages

1. It is necessary to generate a new targeting vector, ES cell line, and mouse strain for each gene to be studied, rather than using pre-existing KO or CKO mice.
2. Typically, after Cre-mediated recombination, the reporter gene is controlled by the promoter and enhancers of the gene that is being deleted, rather than being expressed from a constitutively active promoter like *Rosa26* [21]. Therefore, depending on how the mutant gene is regulated, some mutant cells may not express the reporter gene, so they will not be "visible."

## 5 Mosaic Analysis with Double Markers (MADM)

Mosaic analysis with double markers (MADM) [22] is an ingenious method to generate, in a mouse embryo or tissue, individual mutant cells marked by a fluorescent reporter gene (e.g., GFP) and simultaneously to generate control, WT sister cells marked by a different fluorescent reporter (e.g., Tomato, a red fluorescent protein). The first MADM mice were generated by genetically modifying the *Rosa26* locus on chromosome 6, and they can be used to study only genes on chromosome 6 (and distal to the *Rosa26* locus, with respect to the telomere—as discussed below). For the genes on other mouse chromosome to be amenable to study, it was necessary for the developers of this method to, first, identify a gene on that chromosome, as close as possible to the telomere (*see* **Note 7**), that is expressed ubiquitously (*see* **Note 8**) and then to generate a pair of genetically modified MADM mouse strains with targeted alleles of the gene on that chromosome. MADM strain pairs are currently available (and can be purchased from Jackson Laboratories) for chromosomes 6 [23], 7 [24], 10 [23], 11 [25], and 12 [24].

The MADM design for chromosome 6 is described in Fig. 4a. One mouse strain, MADM6-GT, carries a *Rosa26* allele encoding the 5′ half of a GFP gene and the 3′ half of a Tomato gene, separated by a loxP site. Another strain, MADM6-TG, carries a complementary allele with the 5′ half of Tomato and the 3′ half of GFP. Neither of these alleles encodes a functional fluorescent protein. However, when the two strains are bred together so that all cells carry both a MADM6-TG and a MADM6-GT allele, and Cre recombinase is expressed at a high level, in a very small proportion of cells, there will be an inter-chromosomal recombination event between the loxP site in MADM6-TG and the loxP site in MADM6-GT. This will generate a functional *Rosa26-GFP* gene and a functional *Rosa26-Tomato* gene in the cell (Fig. 4a). When this recombination occurs in the G2 phase of the cell cycle, the chromosome carrying the *Rosa26-GFP* gene can be inherited by one daughter cell and the homologous chromosome 6 carrying the *Rosa26-Tomato* gene inherited by the other daughter cell; thus, one daughter will express GFP (green) and the other Tomato (red). These expression patterns are permanent in the progeny of these two cells. It is important to emphasize that such inter-chromosomal recombination events are very rare, so there are typically very few clones of red/green recombinant cells in a given tissue.

In order to use MADM to generate mosaic embryos or tissues for a KO gene, it is first necessary to position the KO allele on the same chromosome as either the MADM-GT or the MADM-TG allele (but not both). This is usually performed by meiotic recombination, i.e., breeding mice to obtain and identify a meiotic recombination event between the MADM locus and the KO gene of interest (*see* **Note 7**). Once this is accomplished (***and*** if the KO gene is distal to the MADM locus), when Cre recombinase is used to generate a red/green cell pair, the green cell will be homozygous for the KO, and the red cell will be homozygous for the WT allele of that gene (this assumes that the KO gene was placed on the MADM-GT chromosome; if it is placed on the MADM-TG chromosome, then the red cell will be homozygous mutant and the green cell will be homozygous WT).

MADM has been used to generate individual UB cells homozygous for a *Ret*-KO allele on chromosome 6, and sister cells homozygous for the WT *Ret* allele, differentially marked with GFP or Tomato (Fig. 4b) [26]. MADM recombination was induced specifically in UB cells using a Hoxb7/CreGFP transgene [27]. Analysis of developing kidneys with such MADM clones showed that cells lacking *Ret* are much less likely than their WT sister cells to remain at the UB tips during branching morphogenesis [26]. Similar results were observed for the *Etv4* gene on chromosome 11 [26], which functions in the Ret signaling pathway [28].

### 5.1 Materials and Methods

1. Select a Cre-expressing mouse allele that is expressed in the cell type/tissue of interest. Note that MADM apparently does not work with tamoxifen-inducible forms of Cre (e.g., CreER, CreERT2) [20], presumably because the Cre activity is too low to induce recombination between loxP sites on different chromosomes. Theoretically, MADM recombination could also be induced by virally introduced Cre in vitro, but *see* **Note 9**.
2. Obtain, or generate, a KO allele for the gene to be studied, which must be located on one of the chromosomes for which MADM mice are available (*see* above) and must be distal to the MADM locus on that chromosome (*see* **Note 10**).
3. Introduce the KO allele onto the chromosome with one of the two types of MADM loci (e.g., MADM-GT). This is usually done by meiotic recombination (*see* **Notes 7** and **11**).
4. Generate animal colonies that will produce sufficient numbers of the mice needed for **step 5**.
5. Set up crosses that will generate progeny (embryos or live offspring, depending on the stage to be analyzed) that are (a) heterozygous or homozygous for the Cre transgene, (b) heterozygous for the MADM-GT locus *in cis* to the KO allele, and (c) heterozygous for the MADM-TG locus. The expected frequency of progeny that meets all three conditions may be low, but this can be maximized by several methods (*see* **Note 12**).

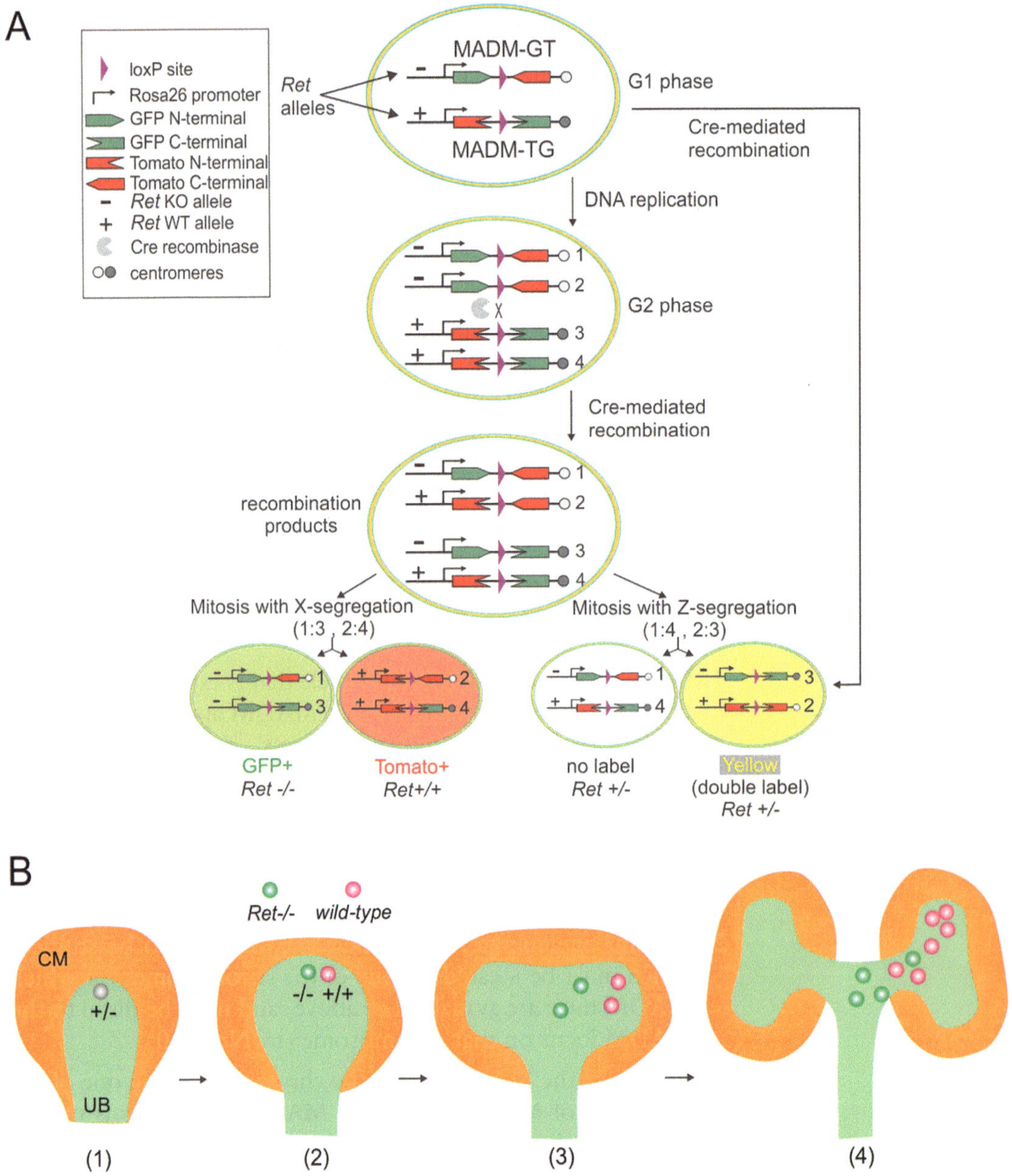


**Fig. 4** Mosaic analysis with double markers (MADM), illustrated for the *Ret* locus on chromosome 6. (**a**) Diagram of MADM recombination. In the initial (G1-phase) cell, one chromosome 6 homolog carries a *Ret*-KO allele *in cis* to *MADM-GT*, and the other carries a *Ret*-WT allele in *cis* to *MADM-TG*. Neither *MADM-GT* nor *MADM-TG* expresses a functional fluorescent protein [22, 23]. Cre-mediated recombination during G1 phase (or during G0, or in post-mitotic cells) generates a Tomato (red fluorescent protein) gene and a GFP gene in the same cell; such cells appear yellow and remain heterozygous for *Ret*. If Cre-mediated recombination occurs after DNA replication, in G2, the cell will contain a Tomato gene in *cis* to a *Ret*-WT allele and a GFP gene in cis to a *Ret*-KO allele, as well as two non-recombined (and nonfunctional) MADM alleles. At mitosis, the four chromatids can undergo X-segregation (left, yielding a *Ret*−/−, GFP+ cell and a *Ret*+/+, Tomato+ cell), or Z-segregation (right, yielding an unlabeled cell and a double-labeled cell, both of which remain *Ret*+/−). The

6. Analyze the properties and behaviors of the green (homozygous KO) vs. the red (homozygous WT) cells in the kidneys of the embryos or mice, e.g., by fluorescence microscopy of cultured kidneys or vibratome sections of larger, more advanced kidneys. The number and size of MADM red/green clones that will be present depend on several factors, including the level of expression of Cre recombinase and the amount of elapsed time during which Cre has been expressed in a given cell lineage. *See* **Notes 13** and **14**.
7. It is sometimes useful to incorporate an additional reporter gene into the cross, to help visualize certain structures in the kidney, such as UB or nephron epithelia (*see* **Note 15**).
8. While MADM clones contain a built-in control (*see* Subheading 5.2), an additional type of control is to generate and analyze MADM red/green clones without any linked KO allele.

### 5.2 Advantages

1. MADM can be performed using any existing KO allele (or other mutant allele, such as a point mutation), as long as it is on a chromosome for which MADM strains are available and distal to the MADM locus on that chromosome.
2. Many Cre lines, expressed in different cell types in the developing kidney, are available to drive recombination in specific cell types.
3. The red/green (WT/mutant) cell pairs contain a built-in control, as the properties of the mutant cells can be directly compared to those of the WT cells.
4. Unlike MASTR clones (*see* below), MADM clones are strongly fluorescent and can be imaged in live or fixed tissue without the need for antibody staining.

### 5.3 Disadvantages

1. MADM mice are not yet available for many chromosomes (although strains for additional chromosomes are being generated).

**Fig. 4** (continued) genotypes and fluorescent protein expression patterns of the initial recombinant cells are retained during subsequent cell divisions. Diagram modified from ref 22, 26. **(b)** Schematic example of the time-lapse analysis of a *Ret*-MADM clone in a cultured kidney. The drawing illustrates a branching UB tip surrounded by cap mesenchyme (CM); the UB expresses Hoxb7/CreGFP (light green), which catalyzes MADM recombination in rare cells. The gray cell in (1) is a parental cell in the *Ret*+/− UB tip, in which Cre-mediated recombination is occurring. After mitosis with X-segregation (2), the *Ret*−/− daughter cell starts to express GFP (darker green), and the *Ret*+/+ cell starts to express Tomato (red). These recombinant cells continue to divide during UB branching (3 and 4), with the red *Ret*+/+ cells tending to remain closer to a tip than the green *Ret*−/− cells [26]

2. The inability to use tamoxifen-inducible forms of Cre means that the timing and extent of recombination cannot be easily controlled, although some control is possible based on the choice of a Cre gene that is expressed at a certain stage of development. But *see* **Note 16**.
3. Breeding strategies to perform MADM can be complex, and a large mouse colony may be required to generate sufficient embryos.

## 6 Mosaic Mutant Analysis with Spatial and Temporal Control of Recombination (MASTR)

As noted above, mosaic kidneys can be generated using any CKO allele, by inducing CreER-mediated recombination in only a subset of cells, but in most cases the recombined (i.e., mutant) cells cannot be distinguished from the WT cells. Unfortunately, it is not possible to use a separate Cre-reporter allele [21, 29] to identify the cells that have recombined the CKO allele, because when CreER is induced at low levels (to generate mosaicism), the correlation between recombination of the CKO and the reporter allele is weak. Lao et al. [30] devised an elegant method that addresses this problem, called mosaic mutant analysis with spatial and temporal control of recombination (MASTR). MASTR makes use of an additional site-specific recombination system, in which FLP recombinase catalyzes recombination between two "frt" recognition sites [31], similar to Cre-mediated recombination between LoxP sites.

As illustrated in Fig. 5, to perform MASTR, mice or embryos are bred to carry three genes: (1) a tamoxifen-inducible Flp gene (such as $Rosa26^{FlpoER}$) [30], (2) the $Rosa26^{MASTR}$ allele [30], and (3) two CKO alleles (or one CKO and one KO) for the gene of interest. The $Rosa26^{MASTR}$ allele includes the *Rosa26* promoter and coding sequences for a GFP-Cre fusion protein, which are preceded by a transcriptional stop sequence flanked by frt sites. Injection of a low dose of tamoxifen activates the FlpoER recombinase, which leads to deletion of the frt-flanked stop sequence in a random subset of cells. These cells then express the GFPcre fusion protein, which simultaneously labels the cells and catalyzes the recombination of the CKO allele(s). The key difference between using MASTR and using a gene encoding CreER (induced by low tamoxifen) to create mosaics is that the Cre activity of the GFPcre fusion protein from the MASTR allele is very strong, so that a CKO allele will be recombined in virtually every cell that expresses the GFPcre. Mosaicism can be achieved by using a low dose of tamoxifen to induce weak FlpoER activity.

Leclerc and Costantini [16] used MASTR to test the importance of fibroblast growth factor receptor 2 (*Fgfr2*) for the ability of

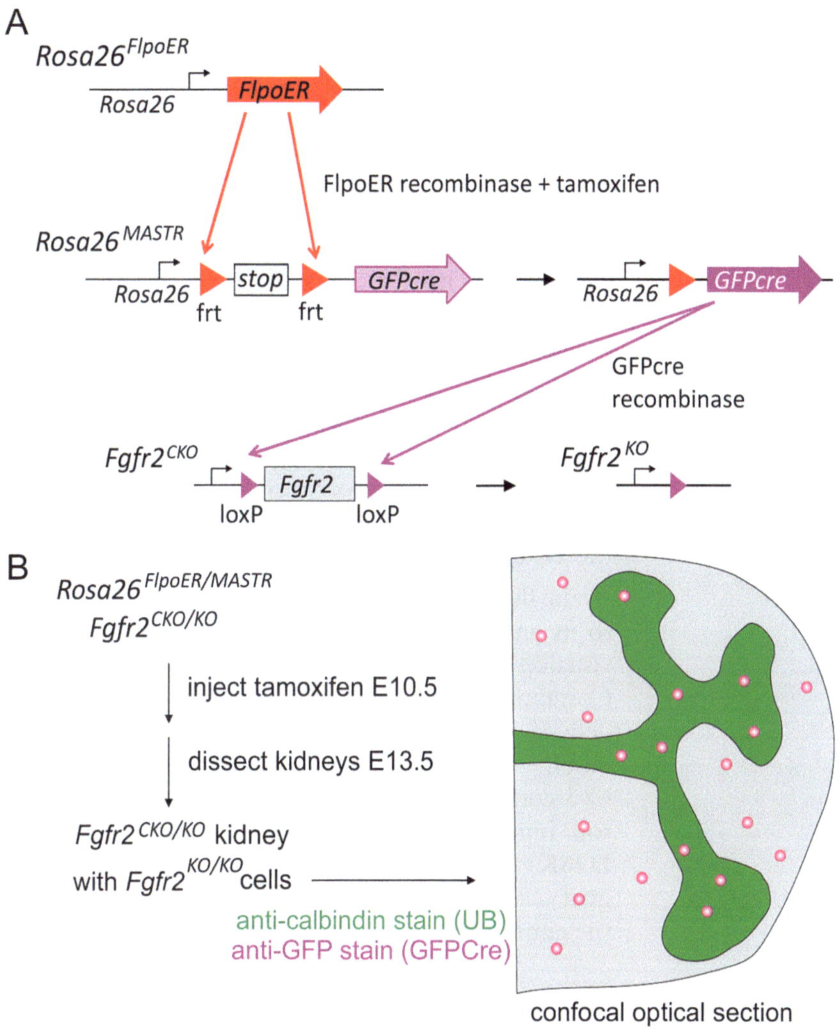


**Fig. 5** Mosaic mutant analysis with spatial and temporal control of recombination (MASTR), illustrated for *Fgfr2*. (**a**) Diagram of MASTR. The *Rosa26*$^{FlpoER}$ allele expresses tamoxifen-inducible FlpoER recombinase. When tamoxifen is injected and binds to FlpoER, FLP activity can induce recombination between frt sites in the *Rosa26*$^{MASTR}$ allele, deleting the stop sequence and causing GFPcre to be expressed. The strong Cre recombinase activity of GFPcre causes an *Fgfr2*$^{CKO}$ allele in the same cell to be converted to a KO allele. (**b**) Use of MASTR to generate *Fgfr2*$^{KO/KO}$ cells in developing kidneys. Note that only a fraction of cells become GFPcre-positive, and delete *Fgfr2*$^{CKO}$ allele. The use of *Rosa26*$^{FlpoER}$ causes recombination in random cells throughout the kidney, which can be detected by staining with anti-GFP. To analyze the effect of *Fgfr2* deletion on UB cells, Leclerc and Costantini [16] analyzed confocal optical sections, where recombinant cells in the UB could be distinguished from those in the mesenchyme, by staining the UB with an anti-Calbindin antibody

UB cells to populate the tip domain during UB branching. Because a UB-specific Flp line was not available (*see* Subheading 6.1.2), a ubiquitously expressed *Rosa26*$^{FlpoER}$ allele [30] was used to delete an *Fgfr2* CKO allele [27] in rare cells throughout the developing kidney. The locations of the CreGFP-expressing recombinant cells were determined by using confocal microscopy to examine optical sections and to count these cells in the UB tips vs. trunks. The results (together with results from the dissociation/aggregation method of generating mosaic renal organoids—*see* Subheading 2) provided evidence that loss of *Fgfr2* reduces the ability of UB cells to remain at the tips [16].

### 6.1 Materials and Methods

1. Set up mating that will yield embryos carrying (1) a tamoxifen-inducible Flp gene such as *Rosa26*$^{FlpoER}$, (2) the *Rosa26*$^{MASTR}$ allele, and (3) two CKO alleles (or one KO and one CKO) of the gene of interest.
2. Inject the pregnant female with tamoxifen at an appropriate stage, typically at least several days before the stage at which the kidneys are to be analyzed. The dose of tamoxifen that yields the desired percentage of recombinant cells must be empirically determined, which can be done using embryos carrying only the *FlpoER* gene and the *Rosa26*$^{MASTR}$ allele.
3. As the fluorescence of GFPcre is weak, it will likely be necessary to fix and stain the kidneys with an antibody against GFP to visualize the recombinant cells (*see* Disadvantages). Leclerc and Costantini [16] fixed whole E13.5 kidneys in 4% paraformaldehyde overnight at 4 °C, washed the kidneys, and stained with rabbit anti-GFP (Invitrogen 1:400) followed by Cy3-conjugated donkey anti-rabbit secondary antibody (Jackson ImmunoResearch). They also used goat anti-Calbindin D28K (Invitrogen, 1:400) and Cy2-conjugated donkey anti-goat secondary antibody (Jackson ImmunoResearch), to stain the entire UB.

### 6.2 Advantages

1. The use of a tamoxifen-inducible FlpoER gene allows the level of mosaicism and the timing of recombination to be controlled by the dose and timing of tamoxifen.
2. MASTR can be performed for any gene for which a CKO allele is available.

### 6.3 Disadvantages

1. Currently, very few mouse lines expressing FlpoER in specific tissues or cell lineages are available. Additional lines should be generated over time.
2. When using *Rosa26*$^{FlpoER}$ together with *Rosa26*$^{MASTR}$, live imaging of cultured kidneys is difficult if not impossible, due to the low fluorescence of GFPcre; only anti-GFP-stained kidneys can be analyzed. However, if a different FlpoER allele (*not* at the *Rosa26* locus) is available, it should be possible to use a *Rosa26*-reporter allele (e.g., *Rosa26*$^{YFP}$) together with *Rosa26*$^{MASTR}$ and the FlpoER allele; GFPcre expression from the *Rosa26*$^{MASTR}$ allele is strong enough to recombine the CKO allele (such as *Fgfr2*$^{CKO}$ in the example cited) as well as

the *Rosa26*-reporter allele in every cell in which *Rosa26*$^{MASTR}$ is activated.

## 7 Notes

1. The host embryos carried the transgene *Hoxb7/Cre* [32] as well as the "reporter allele" *Rosa26R-CFP* [29]; *Hoxb7/Cre* is expressed specifically in the ND and UB, so that Cre-mediated deletion of a floxed transcriptional stop sequence in *Rosa26R-CFP* caused CFP to be expressed permanently in the ND/UB cell lineage.
2. It is desirable to analyze chimeras made with at least two different ES cell lines, in case one of the lines happens to have accidently acquired an unrelated mutation that may affect the results.
3. In some cases ES cell lines that are heterozygous for the KO gene may also be interesting to study, in case heterozygosity also causes phenotypic defects.
4. A potential alternative method to generate mutant ES cells is to use CRISPR/Cas9 transfection [10, 33–35] to knock out both alleles of the gene of interest, in a WT ES cell line that already carries the desired reporter transgene. It may also be possible to transfect the reporter transgene into WT ES cells (rather than deriving a new ES cell line from mice varying the transgene); however, in this case it will be necessary to examine chimeric mice generated from the ES cells, to ensure that the transgene is expressed in the appropriate pattern.
5. Some control over the degree of chimerism can be obtained by injecting different numbers of ES cells per blastocyst. However, even injecting a single ES cell per blastocyst can generate chimeric embryos with >20% mutant cells.
6. It may be difficult to obtain all the necessary embryos, at the same stage of gestation, on the same day. However, it is possible to store E11.5 or E12.5 embryos on ice for up to 3 days, after which the kidneys are still viable [36] and can be used in these experiments.
7. If the KO gene of interest is 5–10 cm or more away from the Rosa26 locus, it is relatively easy to identify a meiotic recombination event that places them on the same chromosome. If it happens to be much closer, the frequency of that meiotic recombination event will be much lower, and it will likely be necessary to breed and screen many more offspring for the recombination event. In such cases, it may be preferable to knock out the gene of interest using CRISPR/Cas9, on the MADM chromosome in MADM6-GT or MADM6-TG mice.

8. A ubiquitously expressed locus to drive expression of the reporter gene is desirable, so that the recombinant cells will always express the reporter, regardless of their fate during subsequent development.
9. Viral vectors (e.g., Adeno-Cre) could be used to introduce Cre recombinase, in organ cultures, for example. However, the virus would need to be microinjected into the area of interest within the kidney, because in kidneys bathed in virus in solution, only the peripheral cells tend to be infected. Another limitation we observed is that infection with Adeno-Cre tended to induce mainly yellow cells (heterozygous for the KO allele) rather than red/green cell pairs, for reasons that are unclear (Cristina Cebrian and F.C., unpublished observations).
10. Only genes that are distal to the MADM locus, with respect to the telomere, will be segregated to generate a homozygous KO and a homozygous WT daughter cell, when recombination between the MADM loci on the two homologous chromosomes is catalyzed by Cre.
11. This is an example for a KO allele located 10 cm distal to the MADM-GT locus on chromosome 6: first, cross a mouse carrying the KO allele with one carrying the MADM6-GT locus. Identify the progeny that inherit both (expected 25% frequency if both parents are heterozygous). In such progeny, the KO allele and the MADM6-GT locus will be located in trans, i.e., each on a different chromosome 6 homolog. Next, cross these progeny to WT mice, and screen the offspring for inheritance of the KO allele and the MADM6-GT locus; since the genes are 10 cm apart, 90% of the offspring will inherit only one (KO allele or MADM locus), 5% will inherit neither, and 5% will inherit both. The latter 5%, by necessity, will be meiotic recombinants carrying both the KO allele and the MADM6-GT locus *in cis*, i.e., on the same chromosome.
12. To maximize the frequency of useful progeny, do one of more of the following:
    (a) Make one parent in the cross homozygous for the Cre gene, so that all offspring will be Cre-positive. Alternatively, make both parents heterozygous for the Cre gene, so 75% of offspring will be Cre-positive.
    (b) Make one parent homozygous for the MADM locus (either MADM-GT or MADM-TG) that is *not* linked *in cis* to the KO allele.
13. Using Hoxb7/CreGFP to induce *Ret*-MADM6 recombination in the UB lineage, Riccio et al. [26] observed that the first MADM clones appeared early in kidney development (e.g., E11.5). The number of new clones increased

exponentially with the growth of the kidney, so that by birth, there were scores if not hundreds of clones per kidney.

14. MADM recombination also generates some cell pairs in which one daughter inherits both the GFP and the Tomato gene and therefore appears yellow by fluorescence microscopy, and the other daughter inherits neither gene and is non-fluorescent (*see* Fig. 4). The yellow cells are heterozygous for the KO gene, like all non-recombinant cells in the embryo, so they can be either ignored or analyzed as control cells.
15. Riccio et al. [26] used two such reporters to label the UB with a fluorescent protein, so that the red/green MADM clones as well as the entire UB were visible by fluorescence microscopy. In one case, they used a Hoxb7/CreGFP transgene [27] to drive recombination of *Ret*-MADM6; the CreGFP fusion protein was weakly fluorescent, rendering the entire UB visible, while the green MADM recombinant cells were more strongly fluorescent. In another case, when performing MADM for *Etv4* on chromosome 11, they included in the embryos a $Rosa26R^{CFP}$ allele [29]; this allele was recombined in most or all UB cells by Hoxb7/Cre [32], making the entire UB CFP-positive.
16. It may be possible to obtain temporal control of MADM recombination, as well as control over the frequency of recombination, by using a different type of inducible system, e.g., control of Cre expression using the Tet system [37, 38].

## Acknowledgments

Research in the author's laboratory has been supported by grants 2R01DK083289 and 5R01DK075578 from the NIH. I thank Dr. Shifaan Thowfeequ for helpful comments on the manuscript.

### References

1. Tsang TE, Shawlot W, Kinder SJ, Kobayashi A, Kwan KM, Schughart K, Kania A, Jessell TM, Behringer RR, Tam PP (2000) Lim1 activity is required for intermediate mesoderm differentiation in the mouse embryo. Dev Biol 223 (1):77–90
2. Schuchardt A, D'Agati V, Pachnis V, Costantini F (1996) Renal agenesis and hypodysplasia in ret-k-mutant mice result from defects in ureteric bud development. Development 122 (6):1919–1929
3. Metzger D, Clifford J, Chiba H, Chambon P (1995) Conditional site-specific recombination in mammalian cells using a ligand-dependent chimeric Cre recombinase. Proc Natl Acad Sci U S A 92(15):6991–6995
4. Kobayashi A, Kwan KM, Carroll TJ, McMahon AP, Mendelsohn CL, Behringer RR (2005) Distinct and sequential tissue-specific activities of the LIM-class homeobox gene Lim1 for tubular morphogenesis during kidney development. Development 132(12):2809–2823
5. McLaren A (1976) Mammalian chimaeras. Cambridge University Press, Cambridge
6. Hogan B, Bedington R, Costantini F, Lacy E (1994) Manipulating the mouse embryo: a laboratory manual, 2nd edn. Cold Spring Harbor Laboratory, Cold Spring Harbor, NY

7. Shakya R, Watanabe T, Costantini F (2005) The role of GDNF/Ret signaling in ureteric bud cell fate and branching morphogenesis. Dev Cell 8(1):65–74
8. Chi X, Michos O, Shakya R, Riccio P, Enomoto H, Licht JD, Asai N, Takahashi M, Ohgami N, Kato M, Mendelsohn C, Costantini F (2009) Ret-dependent cell rearrangements in the Wolffian duct epithelium initiate ureteric bud morphogenesis. Dev Cell 17 (2):199–209
9. Robertson EJ (ed) (1987) Teratocarcinomas and embryonic stem cells: a practical approach, Practical approach series. IRL Press, Oxford
10. Yang H, Wang H, Jaenisch R (2014) Generating genetically modified mice using CRISPR/Cas-mediated genome engineering. Nat Protoc 9(8):1956–1968. https://doi.org/10.1038/nprot.2014.134
11. Costantini F, Watanabe T, Lu B, Chi X, Srinivas S (2011) Imaging kidney development. Cold Spring Harb Protoc 2011(5):468–474. https://doi.org/10.1101/pdb.top109. pii: pdb.top109
12. Lusis M, Li J, Ineson J, Christensen ME, Rice A, Little MH (2010) Isolation of clonogenic, long-term self renewing embryonic renal stem cells. Stem Cell Res 5(1):23–39. https://doi.org/10.1016/j.scr.2010.03.003, pii: S1873-5061(10)00019-X
13. Unbekandt M, Davies JA (2010) Dissociation of embryonic kidneys followed by reaggregation allows the formation of renal tissues. Kidney Int 77(5):407–416. https://doi.org/10.1038/ki.2009.482, pii: ki2009482
14. Ganeva V, Unbekandt M, Davies JA (2011) An improved kidney dissociation and re-aggregation culture system results in nephrons arranged organotypically around a single collecting duct system. Organogenesis 7(2):83–87. https://doi.org/10.4161/org.7.2.14881
15. Xinaris C, Benedetti V, Rizzo P, Abbate M, Corna D, Azzollini N, Conti S, Unbekandt M, Davies JA, Morigi M, Benigni A, Remuzzi G (2012) In vivo maturation of functional renal organoids formed from embryonic cell suspensions. J Am Soc Nephrol 23(11):1857–1868. https://doi.org/10.1681/ASN.2012050505, pii: ASN.2012050505
16. Leclerc K, Costantini F (2016) Mosaic analysis of cell rearrangements during ureteric bud branching in dissociated/reaggregated kidney cultures and in vivo. Dev Dyn 245 (4):483–496. https://doi.org/10.1002/dvdy.24387
17. Davies JA, Unbekandt M, Ineson J, Lusis M, Little MH (2012) Dissociation of embryonic kidney followed by re-aggregation as a method for chimeric analysis. Methods Mol Biol 886:135–146. https://doi.org/10.1007/978-1-61779-851-1_12
18. Uesaka T, Jain S, Yonemura S, Uchiyama Y, Milbrandt J, Enomoto H (2007) Conditional ablation of GFRalpha1 in postmigratory enteric neurons triggers unconventional neuronal death in the colon and causes a Hirschsprung's disease phenotype. Development 134 (11):2171–2181. https://doi.org/10.1242/dev.001388
19. Keefe Davis T, Hoshi M, Jain S (2013) Stage specific requirement of Gfralpha1 in the ureteric epithelium during kidney development. Mech Dev 130(9–10):506–518. https://doi.org/10.1016/j.mod.2013.03.001
20. Feil R, Brocard J, Mascrez B, LeMeur M, Metzger D, Chambon P (1996) Ligand-activated site-specific recombination in mice. Proc Natl Acad Sci U S A 93 (20):10887–10890
21. Soriano P (1999) Generalized lacZ expression with the ROSA26 Cre reporter strain. Nat Genet 21(1):70–71
22. Zong H, Espinosa JS, Su HH, Muzumdar MD, Luo L (2005) Mosaic analysis with double markers in mice. Cell 121(3):479–492
23. Tasic B, Miyamichi K, Hippenmeyer S, Dani VS, Zeng H, Joo W, Zong H, Chen-Tsai Y, Luo L (2012) Extensions of MADM (mosaic analysis with double markers) in mice. PLoS One 7(3):e33332. https://doi.org/10.1371/journal.pone.0033332, pii: PONE-D-11-22207
24. Hippenmeyer S, Johnson RL, Luo L (2013) Mosaic analysis with double markers reveals cell-type-specific paternal growth dominance. Cell Rep 3(3):960–967. https://doi.org/10.1016/j.celrep.2013.02.002, pii: S2211-1247(13)00061-2
25. Hippenmeyer S, Youn YH, Moon HM, Miyamichi K, Zong H, Wynshaw-Boris A, Luo L (2010) Genetic mosaic dissection of Lis1 and Ndel1 in neuronal migration. Neuron 68(4):695–709. https://doi.org/10.1016/j.neuron.2010.09.027, pii: S0896-6273(10)00769-5
26. Riccio P, Cebrian C, Zong H, Hippenmeyer S, Costantini F (2016) Ret and Etv4 promote directed movements of progenitor cells during renal branching morphogenesis. PLoS Biol (in press)
27. Zhao H, Kegg H, Grady S, Truong HT, Robinson ML, Baum M, Bates CM (2004)

Role of fibroblast growth factor receptors 1 and 2 in the ureteric bud. Dev Biol 276 (2):403–415

28. Lu BC, Cebrian C, Chi X, Kuure S, Kuo R, Bates CM, Arber S, Hassell J, MacNeil L, Hoshi M, Jain S, Asai N, Takahashi M, Schmidt-Ott KM, Barasch J, D'Agati V, Costantini F (2009) Etv4 and Etv5 are required downstream of GDNF and Ret for kidney branching morphogenesis. Nat Genet 41 (12):1295–1302. https://doi.org/10.1038/ng.476, pii: ng.476
29. Srinivas S, Watanabe T, Lin CS, William CM, Tanabe Y, Jessell TM, Costantini F (2001) Cre reporter strains produced by targeted insertion of EYFP and ECFP into the ROSA26 locus. BMC Dev Biol 1(1):4
30. Lao Z, Raju GP, Bai CB, Joyner AL (2012) MASTR: a technique for mosaic mutant analysis with spatial and temporal control of recombination using conditional floxed alleles in mice. Cell Rep 2(2):386–396. https://doi.org/10.1016/j.celrep.2012.07.004, pii: S2211-1247(12)00200-8
31. Rodriguez CI, Buchholz F, Galloway J, Sequerra R, Kasper J, Ayala R, Stewart AF, Dymecki SM (2000) High-efficiency deleter mice show that FLPe is an alternative to Cre-loxP. Nat Genet 25(2):139–140
32. Yu J, Carroll TJ, McMahon AP (2002) Sonic hedgehog regulates proliferation and differentiation of mesenchymal cells in the mouse metanephric kidney. Development 129 (22):5301–5312
33. Jinek M, Chylinski K, Fonfara I, Hauer M, Doudna JA, Charpentier E (2012) A programmable dual-RNA-guided DNA endonuclease in adaptive bacterial immunity. Science 337 (6096):816–821. https://doi.org/10.1126/science.1225829
34. Mali P, Yang L, Esvelt KM, Aach J, Guell M, DiCarlo JE, Norville JE, Church GM (2013) RNA-guided human genome engineering via Cas9. Science 339(6121):823–826. https://doi.org/10.1126/science.1232033
35. Cong L, Ran FA, Cox D, Lin S, Barretto R, Habib N, Hsu PD, Wu X, Jiang W, Marraffini LA, Zhang F (2013) Multiplex genome engineering using CRISPR/Cas systems. Science 339(6121):819–823. https://doi.org/10.1126/science.1231143
36. Davies JA (2006) A method for cold storage and transport of viable embryonic kidney rudiments. Kidney Int 70(11):2031–2034
37. Furth PA, St Onge L, Boger H, Gruss P, Gossen M, Kistner A, Bujard H, Hennighausen L (1994) Temporal control of gene expression in transgenic mice by a tetracycline-responsive promoter. Proc Natl Acad Sci U S A 91 (20):9302–9306
38. Gossen M, Freundlieb S, Bender G, Muller G, Hillen W, Bujard H (1995) Transcriptional activation by tetracyclines in mammalian cells. Science 268(5218):1766–1769

# Chapter 2

# Mouse Ex Vivo Kidney Culture Methods


## Anneliis Ihermann-Hella and Satu Kuure



## Abstract

Kidney organogenesis has been a widely used classical model system to study inductive tissue interactions that guide differentiation of many organs. The basis for this is in the pioneering work done during the early 1950s when the conditions of how to support ex vivo growth and differentiation of developing kidneys were revealed. Importantly, culturing developing kidneys remains as an essential instrument to advance our understanding of molecular and cellular regulation of morphogenesis even today. Despite the fact that embryonic kidneys have been cultured for decades, it is not a trivial method and requires specific anatomical and developmental biology knowledge. This chapter outlines the general steps in organ culture and details the requirements for successful kidney explant differentiation.



**Key words** Ex vivo, Organ culture, Kidney development, In vitro differentiation, Ureteric bud, Metanephric mesenchyme, Microdissection


## 1 Introduction

Development of the permanent kidney, the metanephros, begins by the formation of an epithelial, Wolffian duct (WD)-derived ureteric bud (UB). In mouse, which is a widely used model, this takes place around embryonic day 10 (E10) when metanephric mesenchyme (MM), located at the level of anterior end of hind limb, secrets signals that attract WD cells first to cluster and then to bulge on the area where the MM resides [1]. Induction of UB formation is followed by epithelial branching morphogenesis, which begins at E11 when an ampulla of the primary UB is formed [2]. The UB tip initiates its morphogenesis by bifurcating the ampulla and forming the so called T-shaped bud, which is the stage when the first two nephrons are induced [3]. UB tip bifurcations are reiterated until birth, but the majority of the total 10–13 branching rounds take place by E15.5, which after the rate is significantly reduced [4, 5]. Simultaneously to initiation of UB branching, each epithelial tip signals back to the surrounding MM to initiate nephron differentiation [1]. Nephrogenesis is characterized by mesenchyme-to-


Seppo Vainio (ed.), *Kidney Organogenesis: Methods and Protocols*, Methods in Molecular Biology, vol. 1926, https://doi.org/10.1007/978-1-4939-9021-4_2,

epithelium transition (MET), which takes place in the armpits of each T-shaped bud. First the MM condenses and then turns into nephron epithelium through a series of morphologically distinct steps [6]. Therefore UB branching is not only important for the growth and determination of organ shape, but it also instructs the final number of nephrons in the mature organ [7].

Kidneys were one of the earliest organ rudiments that were grown in vitro, and they were extensively used in studies demonstrating that inductive interactions between epithelium and surrounding mesenchyme guide normal differentiation [8]. Kidney ex vivo cultures remain a widely used tool for monitoring morphological events as well as facilitating identification of cellular and molecular regulation leading to functional kidney development [9–12]. Importantly, advances made by utilization of live imaging and lineage tracing in kidney cultures have aspired the use of organ cultures in a wide range of other fields further demonstrating the vital importance of this classical method [13–16].

At least four distinct types of ex vivo kidney cultures exist: Trowel-type air-liquid interphase culture, where tissue is placed on the supportive membrane lying on top of a metal grid or hanging from the plastic frame (commercial inserts) [17]; hanging drop culture, where the organ rudiment is grown on small medium droplets that are maintained by gravity forces [18]; low-volume culture, which utilizes a small silicon ring inside of which organ rudiment is cultured on top of glass slide in a very small amount of culture medium [19]; and fixed Z-dimension culture, where kidney rudiment is squeezed between coverslips supported by a small silicon spacer and supportive membrane [12].The basic Trowel-type kidney culture illustrated in "Dissection and Culture of Mouse Embryonic Kidney" by Aresh and Peuckert published in the *Journal of Visualized Experiments* [20] and described here in details utilizes the same initial steps to dissect organ rudiment from mid-gestation embryos and similar culture medium for the support of their in vitro differentiation.

## 2 Materials

### 2.1 Dissection

1. E11–E12 mouse embryos.
2. Small scissors and fine forceps (*see* **Note 1**).
3. Glass or plastic dishes (100, 60, and 30 mm).
4. Stereomicroscope (e.g., Leica MZ7.5 or Olympus SZX9).
5. Needles (tungsten or 22–25G disposable injection needles with syringes).
6. 50%, 70%, and 95% ethanol (EtOH), diluted in MilliQ-$H_2O$.
7. Dulbecco's PBS (with $Ca^{2+}$ and $Mg^{2+}$) (*see* **Note 2**).
8. 1 M HCl.

### 2.2 Organ Culture

1. Culture medium—DMEM/F12 supplemented with 10% FBS, 1× penicillin-streptomycin, and 1× GlutaMAX.
2. Transwell permeable supports, 24 mm inserts.
3. 6-well CELLSTAR plates.
4. Drummond Wiretrol 10 μL capillaries and delivery plunger.
5. Small pipette tips.
6. Polystyrene beads (corpuscular, 100263-10).
7. Histoacryl® glue (Braun Ref 1050052).
8. Glass coverslips (24 × 24 mm).
9. Microscope equipped with camera.

## 3 Methods

The challenging part in setting up the kidney cultures is recognition and isolation of the rudimentary kidney in E11–E11.5 embryos. These structures are small and don't resemble the mature kidneys people are used to seeing. By E12 the organ rudiment becomes more apparent, and therefore its recognition is easier, and its in vitro differentiation potential improves in comparison with E11 rudiments. Therefore, it is suggested to initiate kidney dissections and cultures with E12.5 kidneys and by this way learn the location and manifestation of the kidney rudiment at its earliest stages (Fig. 1). From the scientific point of view, it should however be kept in mind that same research questions cannot necessarily be studied with E11 and E12.5 kidney rudiments as growing evidence indicates that the MM capping UB tips at E11.5 is distinct from that at E12.5 [10, 21] and the first branching is bifurcation of initial UB into T-bud shape, which differs from the following branching of the T-bud and which also involves trifurcation of at least one of the tips [2, 22]. Therefore, the research question and dissection skills need be taken into account when deciding the stage of the kidney to be cultured (Fig. 2).

### 3.1 Collecting Embryonic Kidneys

1. Euthanize pregnant mouse with $CO_2$ narcotization followed by cervical displacement. Spray the abdomen with 70% EtOH, cut out the uterus with E11–E12.5 embryos, and place it to a 100 mm petri dish filled with Dulbecco's PBS.
2. Stabilize the uterus by holding it still with forceps, and simultaneously open it by incising with scissors.
3. Pull embryos out from the uterus, and rip them free of extra-embryonic tissues with thin forceps (Fig. 1a, b).
4. Use needles to cut the embryo into two halves diagonally just below the liver (Fig. 1c). Remove the hind limbs, liver, and intestine (*see* **Notes 3** and **4**) to get closer to the developing

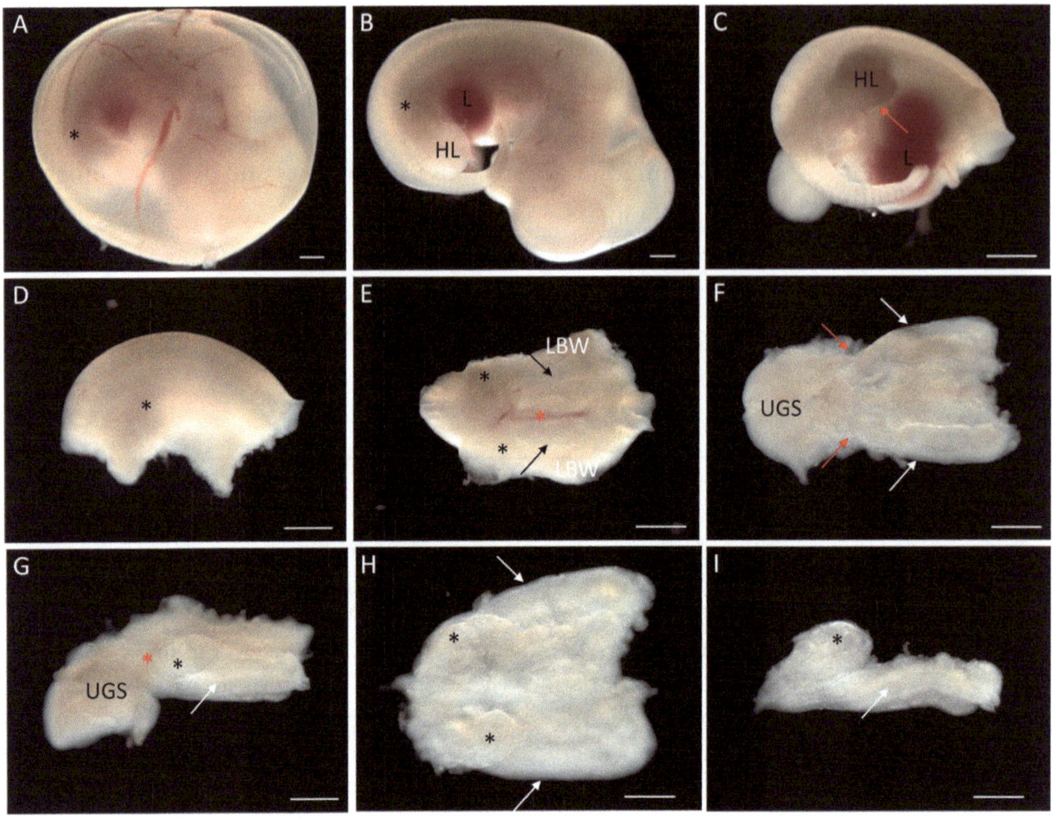


**Fig. 1** Series of images showing different stages of kidney dissection. (**a**) E12.5 mouse embryo inside the amniotic sac. Asterisk indicates an approximate location where developing kidneys are. (**b**) The same embryo is freed from the amniotic sac. *HL* hind limb, *L* liver. (**c**) The dissection is initiated by cutting the embryo just anterior of the liver. Red arrow indicates the developing kidney, which is ventral to the hind limb. (**d**) The same embryo after removing the tail, hind limbs, and internal organs. (**e**) Ventral view of the (**d**), which nicely shows mesonephros and the Wolffian ducts (WD), indicated by black arrows, which laterally surround this transient structure. The dorsal aorta (red asterisk), which is ventral from the spinal cord, is seen as a red stripe in the midline. *LBW* lateral body wall. (**f**) Removal of lateral body walls and dorsal parts of the embryo opens up better sight to the urogenital block. Red arrows point to the developing kidneys, which are hidden behind WDs, and white arrows show WD at the level of mesonephros. *UGS* urogenital sinus. (**g**) Lateral view of specimen in (**f**). Kidney (asterisk) is seen as an oval structure next to the branch of the dorsal aorta (dim red stripe, arteria iliaca externa, red asterisk) going to the limb bud. Wolffian duct (white arrow) is now seen below the developing kidney. (**h**) Dorsal view of the urogenital block after the removal of UGS. Kidneys (asterisks), which show up as upward protrusions, are now located on top and slightly more toward the midline than WDs (white arrows). (**i**) Isolated E12.5 kidney (asterisk) attached to the Wolffian duct (white arrow). Note that anterior (right)-posterior (left) orientation is maintained in all images. Scale bar: (**a**–**e**) 1 mm; (**f**–**i**) 500 μm

kidneys, which are located bilaterally in very close proximity to the spinal cord approximately at the level of the mid-hind legs (Fig. 1d).

5. Start the actual dissection of the kidneys by first removing the dorsal parts of the embryo including the spinal cord. Thereafter turn the embryo to lie on its back, slide the needles into the

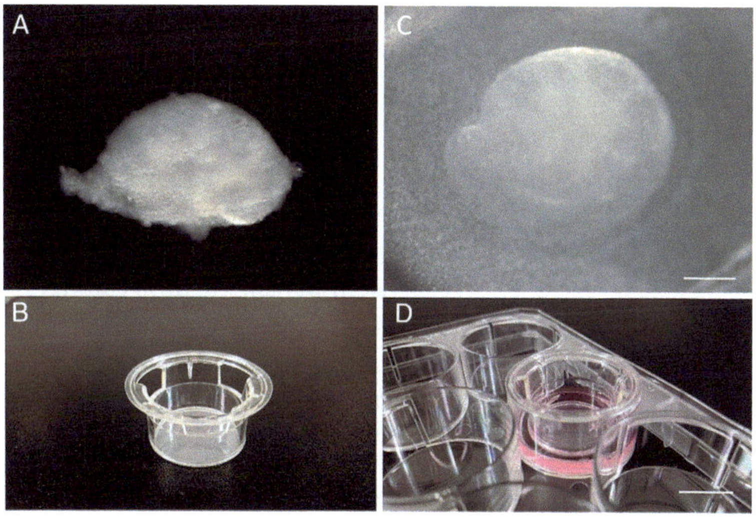


**Fig. 2** Morphology of isolated E12.5 kidney. (**a**) Isolated, intact kidney rudiment. (**b**) Lateral view of Transwell insert. (**c**) E12.5 kidney rudiment cultured for 24 h on Transwell insert, which is in touch with culture medium. Culture of kidney rudiments flattens the organ making it more two-dimensional, and this allows better visualization of UB epithelium (translucent tree-like structure) within the kidney mesenchyme. (**d**) Overall view of Transwell insert in 6-well plate, where kidneys are on top of filter and growth medium underneath. Scale bar (**a**, **c**) 200 μm; (**b**, **d**) 1 cm

lateral body cavities one at a time, and by this way gently cut off the body walls (Fig. 1e–g). Once the kidneys are freed from the embryo, clear off extrarenal tissues as much as possible (Fig. 1h, i, *see* **Note 5**).

### 3.2 Setting up the Trowel-Type of Ex Vivo Kidney Culture

1. Prepare culture dish(es) ready for placing the organ rudiments on Transwell insert filters by adding 1.5 mL of culture medium to the bottom of the dish (*see* **Notes 6** and **7**).
2. Use Wiretrol capillary to place one kidney with as little liquid as possible at the time on its position on the filter (*see* **Note 8**). Remove any visible liquid that surrounds the kidney rudiment by gently sucking the droplet with Wiretrol capillary and avoiding punching the tissue.
3. Place the dish containing kidney rudiments into incubator with +37 °C temperature and 5% $CO_2$.

### 3.3 Setting up the Hanging Drop and Fixed z-Direction Ex Vivo Cultures

For the hanging drop and fixed *z*-direction culture, of which the latter was quite recently published by Saarela and colleagues [12], embryos are collected as described in Subheading 3.1. Thereafter follow the first kidney dissection step in Subheading 3.2, which after the protocols diverge (*see* **Note 9**).

1. To set up hanging drop cultures, maximum of 30 μL medium drops are pipetted on the lid of a petri dish. Pipette the kidney

into each drop with Wiretrol capillary to minimize the amount of medium added. Turn the lid carefully so that the drops start to hang, and place it on top of the dish filled to about 30% with either Dulbecco's PBS or culture medium (*see* **Note 10**). Culture the drops at +37 °C temperature and 5% $CO_2$. For long-term cultures, changing of the drop medium daily may be required due to minimal volume that dries out easily.

2. Fixed *z*-direction culture is reminiscent of Trowel culture with a twist from hanging drop culture. Please *see* Chapter 13 of this book for the details.
   (a) Turn the Transwell insert upside down, and place kidney rudiments on the lower side of it. Add polystyrene beads of 70 μm in diameter to the filter, turn the insert around so that the samples are facing down, and place it against the coverslips in the bottom of the dish (*see* **Note 11**).

### 3.4 Use of Cultured Kidneys

Kidney cultures provide a great tool to observe a living organ and its morphological events. Advancements in microscopy and imaging techniques together with genetic labelling of specific cell types have made live imaging of developing kidney a powerful method significantly advancing the knowledge on cellular events that drive renal organogenesis. At the end of culture period, the kidneys can be used for several purposes. The most used one is to stain the specimens as whole mount samples with tissue and/or cell type-specific markers [23]. This is especially useful when cultured kidneys do not contain genetic labels for distinct cell populations and will significantly augment the depth and diversity of the experiment results. The cultured kidneys can also be used as a source for RNA and protein extraction to be further used in, e.g., expression analyses and Western blotting.

## 4 Notes

1. Use always clean, sterile dissection tools to avoid possible microbial contamination.
2. Instead of Dulbecco's PBS, dissections can be also done in $CO_2$-independent medium.
3. While dissecting the kidneys, keep all other embryos except the one you are working with on ice.
4. E11–E12.5 kidneys are located bilaterally on dorsal midline, ventral to the neural tube and somites, at the level of hind limbs.
5. If studying kidneys of genetically modified mouse lines, store and treat kidneys pairwise from now on, for example, by placing them on 24-well plates.

6. When adding the culture medium to the Transwell insert cultures, pipette it through the small holes in between the insert and dish wall, and make sure not to have air bubbles facing the filter where kidneys will be growing.
7. An alternative for commercial Transwell inserts is the use of metal grids as a support for pieces of nuclepore filters (Whatman, pore size 0.1 μm). Before taking a new batch of nuclepore filters to use, they should be rinsed with distilled $H_2O$ and stored in 70% EtOH. Before using the filters for setting up the culture dish, they need to be well rinsed with distilled $H_2O$ or Dulbecco's PBS, where also smaller pieces can be cut with scissors.
8. When culturing genetically modified kidneys and not knowing the genotype at the time of the culture setup, it is essentially important to place kidneys on filters in a specific order that can be recognized during and after the culture period. Make sure to use markings not only on the lid of the dish but also on its sides to make sure that rotating the lid does not mix the samples.
9. Also metanephric mesenchymes, either in intact or dissociated form, and ureteric buds (*see* Chapters 4 and 5 for details on how to get these) can be cultured in hanging drop system.
10. It is important not to move the lid sideways while turning it upside down to avoid spreading of drops where kidneys are placed.
11. It is advised to fix the whole insert into one position on the dish by melting its rim with a heated glass capillary at three points.

## References


1. Saxen L (1987) Organogenesis of the kidney. Cambridge University Press, Cambridge
2. Costantini F (2012) Genetic controls and cellular behaviors in branching morphogenesis of the renal collecting system. Wiley Interdiscip Rev Dev Biol 1:693–713
3. O'Brien LL, McMahon AP (2014) Induction and patterning of the metanephric nephron. Semin Cell Dev Biol 36:31–38
4. Cebrian C, Borodo K, Charles N, Herzlinger DA (2004) Morphometric index of the developing murine kidney. Dev Dyn 231:601–608
5. Short KM, Combes AN, Lefevre J, Ju AL, Georgas KM, Lamberton T, Cairncross O, Rumballe BA, McMahon AP, Hamilton NA et al (2014) Global quantification of tissue dynamics in the developing mouse kidney. Dev Cell 29:188–202
6. Quaggin SE, Kreidberg JA (2008) Development of the renal glomerulus: good neighbors and good fences. Development 135:609–620
7. Schedl A (2007) Renal abnormalities and their developmental origin. Nat Rev Genet 8:791–802
8. Grobstein C (1953) Inductive epitheliomesenchymal interaction in cultured organ rudiments of the mouse. Science 118:52–55
9. Motamedi FJ, Badro DA, Clarkson M, Lecca MR, Bradford ST, Buske FA, Saar K, Hubner N, Brandli AW, Schedl A (2014) WT1 controls antagonistic FGF and BMP-pSMAD pathways in early renal progenitors. Nat Commun 5:4444
10. Ihermann-Hella A, Lume M, Miinalainen IJ, Pirttiniemi A, Gui Y, Peranen J, Charron J, Saarma M, Costantini F, Kuure S (2014) Mitogen-activated protein kinase (MAPK)

pathway regulates branching by remodeling epithelial cell adhesion. PLoS Genet 10: e1004193

11. Riccio P, Cebrian C, Zong H, Hippenmeyer S, Costantini F (2016) Ret and Etv4 promote directed movements of progenitor cells during renal branching morphogenesis. PLoS Biol 14: e1002382
12. Saarela U, Akram SU, Desgrange A, Rak-Raszewska A, Shan J, Cereghini S, Ronkainen VP, Heikkila J, Skovorodkin I, Vainio SJ (2017) Novel fixed z-direction (FiZD) kidney primordia and an organoid culture system for time-lapse confocal imaging. Development 144:1113–1117
13. Voutilainen M, Lindfors PH, Lefebvre S, Ahtiainen L, Fliniaux I, Rysti E, Murtoniemi M, Schneider P, Schmidt-Ullrich-R, Mikkola ML (2012) Ectodysplasin regulates hormone-independent mammary ductal morphogenesis via NF-kappaB. Proc Natl Acad Sci U S A 109:5744–5749
14. Speroni L, Voutilainen M, Mikkola ML, Klager SA, Schaeberle CM, Sonnenschein C, Soto AM (2017) New insights into fetal mammary gland morphogenesis: differential effects of natural and environmental estrogens. Sci Rep 7:40806
15. Burke ZD, Li WC, Slack JM, Tosh D (2010) Isolation and culture of embryonic pancreas and liver. Methods Mol Biol 633:91–99
16. Shih HP, Sander M (2014) Pancreas development ex vivo: culturing embryonic pancreas explants on permeable culture inserts, with fibronectin-coated glass microwells, or embedded in three-dimensional matrigel. Methods Mol Biol 1210:229–237
17. Trowell OA (1954) A modified technique for organ culture in vitro. Exp Cell Res 6:246–248
18. Saxen L, Lehtonen E (1987) Embryonic kidney in organ culture. Differentiation 36:2–11
19. Sebinger DD, Unbekandt M, Ganeva VV, Ofenbauer A, Werner C, Davies JA (2010) A novel, low-volume method for organ culture of embryonic kidneys that allows development of cortico-medullary anatomical organization. PLoS One 5:e10550
20. Aresh B, Peuckert C (2017) Dissection and culture of mouse embryonic kidney. J Vis Exp (123):55715
21. Nie X, Xu J, El-Hashash A, Xu PX (2011) Six1 regulates Grem1 expression in the metanephric mesenchyme to initiate branching morphogenesis. Dev Biol 352:141–151
22. Majumdar A, Vainio S, Kispert A, McMahon J, McMahon AP (2003) Wnt11 and Ret/Gdnf pathways cooperate in regulating ureteric branching during metanephric kidney development. Development 130:3175–3185
23. Kuure S, Sainio K, Vuolteenaho R, Ilves M, Wartiovaara K, Immonen T, Kvist J, Vainio S, Sariola H (2005) Crosstalk between Jagged1 and GDNF/Ret/GFRalpha1 signalling regulates ureteric budding and branching. Mech Dev 122:765–780

# Chapter 3

# Serum-Free Organ Culture of the Embryonic Mouse Ureter

**Filipa M. Lopes and Adrian S. Woolf**


## Abstract

The ability to explant and then maintain embryonic tissues in organ culture makes it feasible to study the growth and differentiation of whole organs, or parts or combinations of them, in three dimensions. Moreover, the possible effects of biochemical manipulations or mutations can be explored by visualizing a growing organ. The mammalian renal tract comprises the kidney, ureter, and urinary bladder, and the focus of this chapter is organ culture of the embryonic mouse ureter in serum-free defined medium. Over the culture period, rudiments grow in length, smooth muscle differentiates, and the ureters then undergo peristalsis in a proximal to distal direction.


**Key words** Culture, Embryo, Mouse, Peristalsis, Ureter

## 1 Introduction

The mammalian renal tract comprises the kidney, ureter, and urinary bladder, and the focus of this chapter is organ culture of the embryonic mouse ureter in serum-free defined medium. The ability to explant and then maintain embryonic tissues in organ culture makes it feasible to directly visualize and measure the growth and differentiation of individual organs, or parts or combinations of them. The methodological foundations of mammalian organ culture took place in the 1950s when pioneers described the differentiation of explants on clots [1] or on grids at the interface of air and medium [2]. In subsequent decades embryonic kidney organ culture was used to determine effects on growth and differentiation of, as examples, blocking DNA-dependent RNA synthesis [3], perturbing extracellular matrix molecules [4], or adding molecules affecting tubular physiology [5]. An important refinement to the organ culture technique was the use of serum-free defined medium supplemented with insulin, selenium, and transferrin to nurture growth and differentiation, and this has been applied to explants of embryonic kidneys [6], urinary bladders [7], and ureters

Seppo Vainio (ed.), *Kidney Organogenesis: Methods and Protocols*, Methods in Molecular Biology, vol. 1926,
https://doi.org/10.1007/978-1-4939-9021-4_3, 

[8]. This advance avoided poorly defined, and hard to reproduce, biological factors such as chick embryo juice and horse serum [1].

The mature ureter is a cylindrical tapering tube that connects the pelvis of the kidney with the urinary bladder. Its lumen is lined by specialized epithelia called urothelia, and this is enveloped by smooth muscle. Other components include an elastic fibroblastic layer between the urothelium and muscle and Cajal-like cells within the muscle layer that are required for the transmission of proximal (top of the ureter) to distal (bottom of the ureter) peristaltic waves that are intrinsic to the organ [9] and which propel urine from the renal pelvis to the bladder [8]. The peristaltic waves originate in a "pacemaker" in the junction of the ureter and renal pelvis [10, 11]. The mouse ureter initiates at embryonic day 10.5, when it consists of the ureteric bud, the epithelial branch originating in the distal region of the pronephric duct [12]. Over the next prenatal week in vivo, the ureteric bud elongates to form the urothelial core of the stalk of the ureter, while the top of the bud interacts with metanephric mesenchyme and undergoes recurrent branching to form the kidney collecting ducts. The junction of the stalk and the collecting ducts remodel to form the renal pelvis. Meanwhile, mesenchymal cells condense around the epithelial stalk and then differentiate into smooth muscle cells. Several of the molecular drivers of these events have been identified; their detailed description is beyond the remint of this chapter but can be found in other refs. 8, 10, 12–14. In humans, malformations of the ureter are among the commonest congenital anomalies and are often detected upon prenatal ultrasound screening [15]. They can arise as secondary events, for example, from physical obstruction of the bladder outlet or from mutations of genes that control ureteric growth and differentiation [16, 17].

## 2 Materials

Unless otherwise stated, all chemicals were obtained from Sigma-Aldrich.

### 2.1 Mice

CD1 outbred wild-type mice were maintained in a colony housed in the Biological Service Facility of the University of Manchester. CD1 mice are also commercially obtainable from, for example, Charles River.

### 2.2 Materials

1. Ethanol, 70%.
2. Forceps.
3. Glass Petri dish.
4. Ice.
5. Needles, 20G; Leibovitz's L-15 medium.

6. Plastic Petri dishes.
7. Small scissors.
8. Inverted light microscope (Leica M80).
9. Serum-free culture medium:
   (a) Dulbecco's Modified Eagle Medium: nutrient mixture F-12 (DMEM/F-12) with L-glutamine, 15 mM HEPES.
   (b) Sodium bicarbonate insulin-transferrin-selenium (ITS) supplement, ×100 (Life Technologies).
   (c) Penicillin-streptomycin (P/S) antibiotic, ×100 (Thermo Fisher Scientific).

   All the components should be pre-warmed at 37 °C.
10. Laminar flow hood, class II safety cabinet (UniMat-2).
11. Millicell culture plate semipermeable inserts, 30 mm in diameter, made from polytetrafluoroethylene with 0.4 μm pores (Millipore).
12. Pipettes and tips (2–1000 μL).
13. Plastic six-well plate dish.
14. Tissue culture incubator.
15. Inverted microscope Nikon eclipse Ti-E with time lapse equipped with chamber able to maintain 37 °C, 5% $CO_2$.

## 3 Methods

The procedures are undertaken in an aseptic, laboratory environment. Preparation of medium and feeding of organ cultures are undertaken in a laminar flow hood.

### 3.1 Embryonic Ureter Dissection

Figure 1a depicts a flowchart of the procedures used to isolate the ureter. Figure 1b depicts a whole E15 embryo, Fig. 1c shows the whole renal tract as it appears within the embryo, and Fig. 1d shows the renal tract at the penultimate stage of the dissection when the embryonic kidney is still attached to the ureter (*see* **Note 1**).

1. Mice were mated overnight. The next morning, when the vaginal plug was detected, is designated as "embryonic day 0" (E0) (*see* **Note 2**).
2. The time-mated pregnant mouse is euthanized on the morning when embryos are E15. Sterilize the abdomen with 70% ethanol. Make a "V"-shaped incision in the abdomen, and remove the uterine horns containing the embryos. Place the uterine horn in L-15 medium, and isolate each embryo from its enveloping membranes. Place embryos into L-15 medium in a plastic Petri dish cooled on ice.

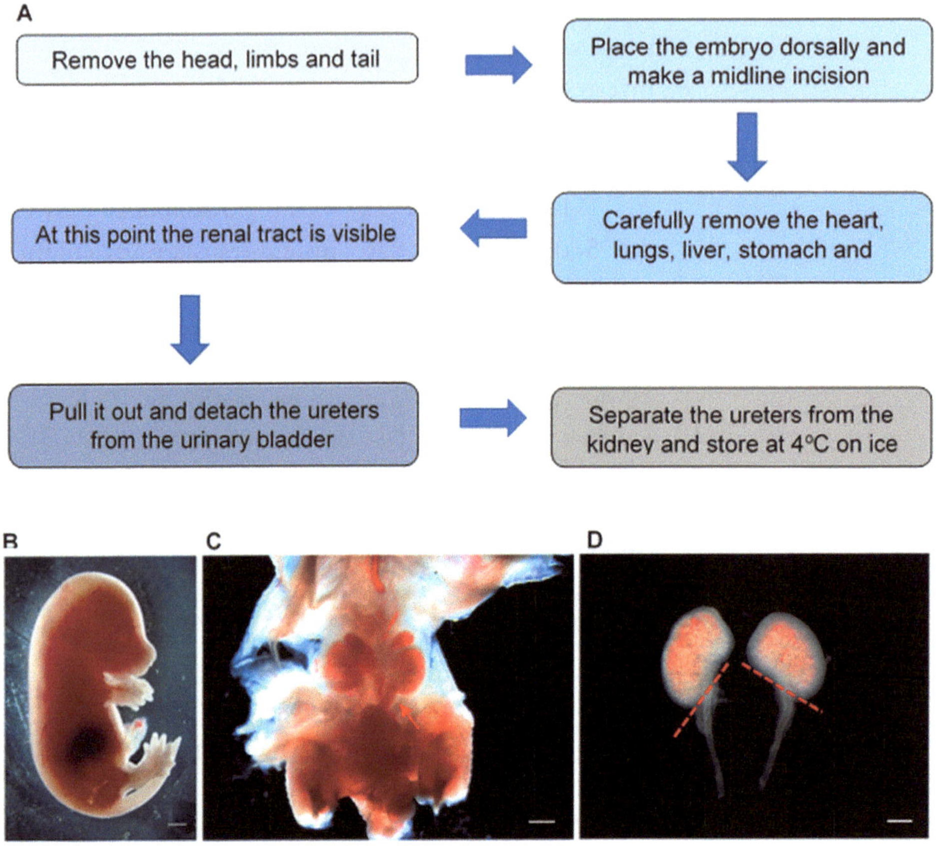


**Fig. 1** Isolation of the embryonic mouse ureter. (**a**) Depicts a flowchart of the procedures used to isolate the ureter. (**b**) Depicts a whole E15 embryo. (**c**) Shows the intact tract as it appears in the dissected embryo: arrow indicates the left ureter, with the metanephric kidney above and the urinary bladder below. (**d**) Shows the isolated renal tract at the penultimate stage of the dissection when the embryonic metanephric kidney is still attached to the ureter: the red dotted lines indicate the plane of section used to separate the ureter from the kidney

3. Under the dissecting microscope, dissect each embryo in a separate dish. Use forceps to remove the head and limbs. Make a midline incision on the front of the abdomen. Gently, pull out all the organs up until the renal tract is reached. Taking the kidney as a reference point, use a needle to detach the ureter from the bladder by gently pulling on the kidney (*see* **Note 3**). Cut the ureter where it enters the embryonic kidney. This is the top/proximal portion of the ureter and is wider than the bottom/distal end.
4. Using a 200 μL pipette, selected because the tip of this pipette is wide enough to accommodate the rudiment, transfer the embryonic ureters in L-15 medium to a clean Petri dish and maintain it at 4 °C.

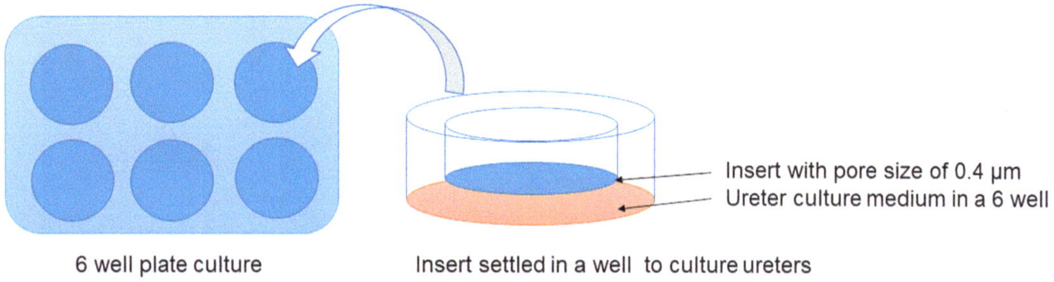


**Fig. 2** Diagram of the organ culture insert. Each insert is placed inside a well within the six-well plate. Medium is added so that it reaches and wets the bottom of the insert

### *3.2 Embryonic Ureter Culture*

1. Add 1.20 mL of serum-free medium per well to be used of the six-well plate, and place one Millicell culture plate insert into each well (*see* **Note 4**). The filter will become wetted by the medium, forming a platform exposed to the humidified atmosphere (Fig. 2). The platforms provide a solid base on which organs can differentiate and grow, while accessing chemical nutrients from medium below the platform.
2. Use a 200 μL pipette to transfer the embryonic ureters onto the surface of the platform.
3. Place the plastic cover over the wells, and place the six-well plate into the tissue culture incubator maintained at 37 °C in a humidified air/5% $CO_2$ atmosphere.
4. After 3 days in culture, replace the medium with fresh medium.
5. Make and record daily microscopic observations to monitor the ureteric growth. The plate is taken out from the incubator and transported to the microscopic room only during the necessary time to observe and take pictures. Measure the length and area on each day of culture using ImageJ freehand tool. Figure 3 depicts the growth of the explanted ureter over 6 days in culture and its change in terms of gross morphology using the above protocol. By the end of the culture period, a smooth muscle layer is seen to surround the epithelial core of the explant. At this stage, the ureter is spontaneously contractile.
6. At the last day of culture, take out the six-well plate culture from the incubator, and place it under a microscope equipped with a chamber able to maintain the physiologic conditions for ureter survival (37 °C, 5% $CO_2$), allowing them to acclimatize for 20 min.
7. Open the NIS-Elements Advanced Research software linked to the microscope, select the live option, and tally the number of contractions during 2 min. If various groups are being tested, compare between them. Time-lapse imaging option is also performed to record ureteric contractions for 2 min in culture using this same microscope. The image acquisition is made every 0.2 s using bright-field light and a 4× objective.

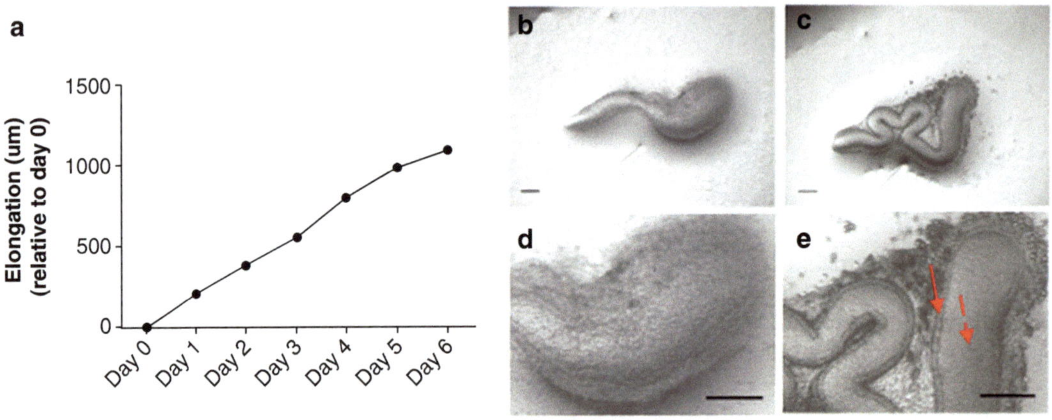


**Fig. 3** Ureter growth and differentiation over 6 days in organ culture. (**a**) Graph depicting linear growth of the explanted E15 mouse ureter over 6 days in organ culture. (**b**, **d**) They are low- and high-power photographs taken through an inverted microscope showing the simple morphology of the E15 ureter at day 0 of culture. (**c**, **e**) They are low- and high-power images of the same explant which exemplify at day 6 of organ culture. The core of the explant, indicated by the dashed red arrow, is the epithelium. This is surrounded by another zone, the smooth muscle layer, indicted by the solid red arrow. Scale bars are 200 μm

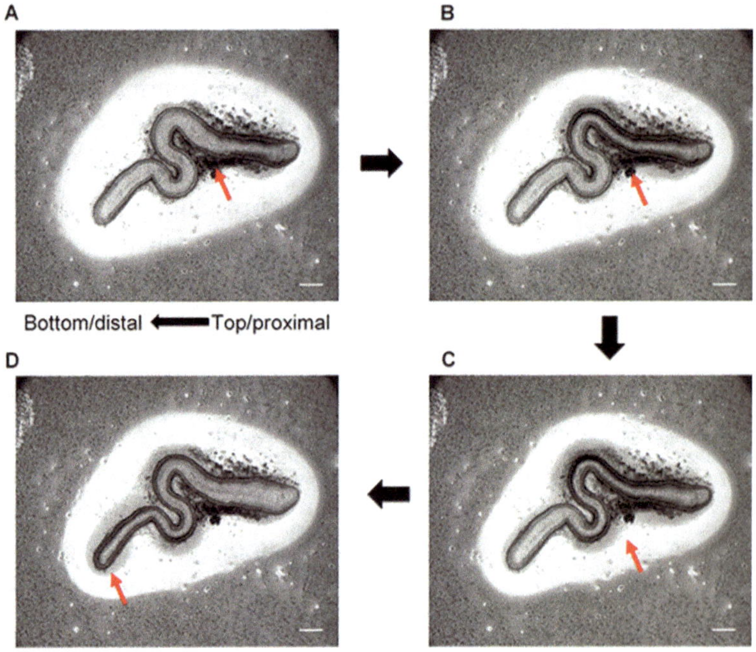


**Fig. 4** Spontaneous peristalsis in the ureter at day 6 of organ culture. The contraction is a wavelike movement that starts at the top/proximal (close to the kidney) end (**a**) of the ureter and propagates (**b**, **c**) toward the bottom/distal (urinary bladder) end (**d**) of the ureter. In vivo, this mechanism propels urine from the pelvis of the kidney to the bladder. Scale bar is 200 μm

8. The images obtained are converted to an AVI (Audio Video Interleaved) format using the ImageJ and played at 5.3 frames/s. As depicted in Fig. 4, contractions originate near the top, or proximal end, of the ureter, and a subset of these contractions are perpetuated as peristaltic waves that move down the explanted organ to its bottom, or distal, end. Therefore, this is a robust model with which is possible to study the elongation, morphology, and function of the ureter during embryonic development.

## 4 Notes

1. The protocol described is designed for E15 mouse ureters. The protocol can be used to compare mutant with wild-type embryonic ureters [8]. The current methodology can also be applied to culture of the embryonic kidney [5], urinary bladder culture [7], and small intestine [18].
2. The use of mice in these experiments must be approved by local and, when required, national ethics committees.
3. If the ureter does not detach from the urinary bladder, cut it using the needle, in a way to get as much ureter as possible.
4. If not all the six wells are being used, add 1.20 mL of medium to the empty ones, so that the environment is humid and the most homogenous possible over the plate.

## Acknowledgments

This work was supported by the RENALTRACT Initial Training Network that received funding from the European Union's Horizon 2020 research and innovation program under the Marie Sklodowska-Curie grant agreement no. 642937.

### References

1. Grobstein C (1953) Inductive epitheliomesenchymal interaction in cultured organ rudiments of the mouse. Science 118:52–55
2. Trowell OA (1954) A modified technique for organ culture in vitro. Exp Cell Res 6:246–248
3. Jainchill J, Saxen L, Vainio T (1964) Studies on kidney tubulogenesis. 1. The effect of actinomycin D on tubulogenesis in vitro. J Embryol Exp Morphol 12:597–607
4. Klein G, Langegger M, Timpl R, Ekblom P (1988) Role of laminin A chain in the development of epithelial cell polarity. Cell 55:331–341
5. Chan SK, Riley PR, Price KL, McElduff F, Winyard PJ, Welham SJ, Woolf AS, Long DA (2010) Corticosteroid-induced kidney dysmorphogenesis is associated with deregulated expression of known cystogenic molecules, as well as Indian hedgehog. Am J Physiol Renal Physiol 298:F346–F356
6. Avner ED, Jaffe R, Temple T, Ellis D, Chung AE (1983) Development of renal basement membrane glycoproteins in metanephric organ culture. Lab Investig 48:263–268
7. Burgu B, McCarthy LS, Shah V, Long DA, Wilcox DT, Woolf AS (2006) Vascular

endothelial growth factor stimulates embryonic urinary bladder development in organ culture. BJU Int 98:217–225
8. Caubit X, Lye CM, Martin E, Coré N, Long DA, Vola C, Jenkins D, Garratt AN, Skaer H, Woolf AS, Fasano L (2008) Teashirt 3 is necessary for ureteral smooth muscle differentiation downstream of SHH and BMP4. Development 135:3301–3310
9. David SG, Cebrian C, Vaughan ED Jr, Herzlinger D (2005) c-Kit and ureteral peristalsis. J Urol 173:292–295
10. Cain JE, Islam E, Haxho F, Blake J, Rosenblum ND (2011) GLI3 repressor controls functional development of the mouse ureter. J Clin Invest 121:1199–1206
11. Hurtado R, Bub G, Herzlinger D (2014) A molecular signature of tissues with pacemaker activity in the heart and upper urinary tract involves coexpressed hyperpolarization-activated cation and T-type Ca2+ channels. FASEB J 28:730–739
12. Woolf AS, Davies JA (2013) Cell biology of ureter development. J Am Soc Nephrol 24:19–25
13. Bohnenpoll T, Wittern AB, Mamo TM, Weiss AC, Rudat C, Kleppa MJ, Schuster-Gossler K, Wojahn I, Lüdtke TH, Trowe MO, Kispert A (2017) A SHH-FOXF1-BMP4 signaling axis regulating growth and differentiation of epithelial and mesenchymal tissues in ureter development. PLoS Genet 13:e1006951
14. Pitera JE, Woolf AS, Basson AM, Scambler PJ (2012) *Sprouty1* haploinsufficiency prevents renal agenesis in a model of Fraser syndrome. J Am Soc Nephrol 23:1790–1796
15. Kerecuk L, Schreuder MF, Woolf AS (2008) Renal tract malformations: perspectives for nephrologists. Nat Clin Pract Nephrol 4:312–325
16. Caubit X, Gubellini P, Andrieux J, Roubertoux P, Metwaly M, Jacq B, Fatmi A, Had-Aissoini L, Kwan K, Salin P, Carlier M, Liedén A, Rudd E, Shinawi M, Vincent-Delorme C, Cuisset J-M, Lemaitre M-P, Abderrehamane F, Duban B, Lemaitre J-F, Woolf AS, Bockenhauer D, Severac D, Dubois E, Zhu Y, Šestan N, Garratt AN, Kerkerian-Le Goff L, Fasano L (2016) *TSHZ3* deletion causes an autism syndrome and defects in cortical projection neurons. Nat Genet 48:1359–1369
17. Cuvertino S, Stuart HM, Chandler KE, Roberts NA, Armstrong R, Bernardini L, Bhaskar S, Callewaert B, Clayton-Smith J, Hernando-Davalillo C, Deshpande C, Devriendt K, Digilio MC, Dixit A, Edwards M, Friedman JM, Gonzalez-Meneses-A, Joss S, Kerr B, Lampe AK, Langlois SL, Loget P, McGowan R, De Smedt M, O'Sullivan J, Odent S, Parker MJ, Pebrel-Richard C, Petit F, Stark Z, Tinschert S, Vasudevan P, Villa O, White SM, Zahir FR, Lennon R, The DDD Study, Woolf AS, Banka S (2017) *ACTB* loss-of-function mutations result in a pleiotropic developmental disorder. Am J Hum Genet 101:1021–1033
18. Coletta R, Roberts NA, Oltrabella F, Khalil BA, Morabito A, Woolf AS (2016) Bridging the gap: functional healing of embryonic small intestine ex vivo. J Tissue Eng Regen Med 10:178–182

Chapter 4

# Experimental Tubulogenesis Induction Model in the Mouse

Aleksandra Rak-Raszewska

## Abstract

Kidney development and induction of tubulogenesis have been studied for almost seven decades. The experimental setup of metanephric mesenchyme induction ex vivo allows to control the environment, to perform cellular manipulations, and to learn about renal development. Since the establishment of the ex vivo kidney culture technique in 1953, the method was modified to suit well the progress in biological and medical fields and still today present many advantages over the traditional in vivo studies.

**Key words** Metanephric mesenchyme induction, Tubulogenesis, Nephrogenesis, Ureteric bud, Spinal cord, BIO, LiCl

## 1 Introduction

The permanent (metanephric) kidneys develop from the intermediate mesoderm, which gives rise to ureteric bud (UB) and metanephric mesenchyme (MM). Reciprocal interactions of these two tissues initiate nephrogenesis which leads to development of functional units of the kidney—the nephrons [1]. The first experimental nephrogenesis induction was performed by Grobstein [2, 3], and his technique was modified later by Trowell [4] and Saxen [5]. Grobstein showed also that neither MM nor UB survives when they are not in contact with each other [2] and that MM can be induced with the use of embryonic spinal cord [6]. Recently, this method came to a spotlight again, when Unbekandt and Davies have proposed dissociation and reaggregation of embryonic kidneys in order to enable more efficient manipulation of cells [7].

It has been shown that on molecular level, the induction signal to MM is led by activation of Wnt signaling pathway [8] and later that it is possible to replace it with chemical inducers such as the lithium chloride [9, 10] or BIO (6-bromoindirubin-3′-oxime) [11]. Both chemicals inactivate the GSK3 in the "destruction complex" and allow the nuclear transfer of the β-catenin leading to activation of target genes [11]. The experimental tubulogenesis

Seppo Vainio (ed.), *Kidney Organogenesis: Methods and Protocols*, Methods in Molecular Biology, vol. 1926,
https://doi.org/10.1007/978-1-4939-9021-4_4, 

induction model has evolved in recent years, and due to significant increase in efficiency of cellular manipulations, the dissociated MM is used more often nowadays [12]. The induction may also be directed by application of whole UB or dissociated UB but also can be replaced with embryonic spinal cord or chemicals. Each modification is presenting different advantages and specific uses; they are presented here.

## 2 Materials

### 2.1 Dissociation of Embryonic Kidneys

1. Dissected embryonic kidneys in PBS−/− on ice (*see* **Note 1**).
2. Simple light microscope, centrifuge, 15 mL falcon tube, pipettes, pipette's tips, 40 μm filters, water bath set to temp. 37 °C.
3. Complete kidney culture medium: 10% FCS, 1% penicillin/streptomycin, DMEM high glucose (*see* **Note 2**). Filter the medium using 0.22 μm syringe filters (*see* **Note 3**).

### 2.2 Separation of the Ureteric Bud from the Metanephric Mesenchyme

1. Dissected embryonic kidneys in PBS−/− on ice (*see* **Note 1**).
2. Trypsin/pancreatin solution: It is a mixture of two reagents, trypsin and pancreatin, in Tyrode's solution. Measure: 0.450 g of trypsin (Sigma), 2 mL of pancreatin (*see* **Note 4**), 18 mL of Tyrode's solution (*see* **Note 5**). Dissolve trypsin on ice and then measure pH; it should be 7.4 (*see* **Note 6**).
3. Complete kidney culture medium: 10% FCS, 1% penicillin/streptomycin, DMEM high glucose (*see* **Note 2**). Filter the medium using 0.22 μm syringe filters (*see* **Note 3**).
4. Dissecting microscope, needle size G27, glass dish, pipettes, and pipette's tips.

### 2.3 Dissociation of the Metanephric Mesenchyme

1. Metanephric mesenchyme devoid of ureteric buds.
2. Dissociation solution: Measure 280 μL of 0.5% BSA in PBS (−/−) (*see* **Note 7**) and 40 μL of collagenase III or IV (*see* **Note 8**).
3. Complete kidney culture medium: 10% FCS, 1% penicillin/streptomycin, DMEM high glucose (*see* **Note 2**). Filter the medium using 0.22 μm syringe filters (*see* **Note 3**).
4. Dissecting microscope, low binding Eppendorf tube, pipettes and pipette's tips, incubator or water bath.

### 2.4 Dissociation of Ureteric Buds

1. Ureteric buds devoid of metanephric mesenchyme.
2. 1× trypsin/EDTA solution.

3. Complete kidney culture medium: 10% FCS, 1% penicillin/streptomycin, DMEM high glucose (*see* **Note 2**). Filter the medium using 0.22 μm syringe filters (*see* **Note 3**).
4. Dissecting microscope, Eppendorf tube, pipettes and pipette's tips, incubator or water bath.

### 2.5 Dissection of Embryonic Spinal Cord

1. Cranial part of the mouse E11.5 embryo.
2. Dissecting microscope, needle size G27, PBS−/−.

### 2.6 Tubulogenesis Induction

1. Dissected tissues (MM, UB) as required.
2. Complete kidney culture medium: 10% FCS, 1% penicillin/streptomycin, DMEM high glucose (*see* **Note 2**). Filter the medium using 0.22 μm syringe filters (*see* **Note 3**).
3. Chemicals: BIO (6-bromoindirubin-3′-oxime) or LiCl.
4. Dissecting microscope, pipettes and pipette's tips, 6-well or 12-well Transwell filter dish.

## 3 Methods

### 3.1 Dissociation of Embryonic Kidneys

1. Place your E11.5 kidneys into 15 mL falcon tube with 3 mL of 1× trypsin/EDTA solution (*see* **Note 9**).
2. Incubate in water bath at 37 °C for max. of 10 min (*see* **Note 10**).
3. Pipette vigorously every 5 min (*see* **Note 11**).
4. Once you can identify single cells down the microscope and they are in majority, stop the enzymatic reaction with 7 mL of complete kidney culture medium (*see* Subheading 2.1 and **Note 2**). Mix well using 1 mL pipette (*see* **Note 11**).
5. Incubate cells in the water bath at 37 °C for further 10 min.
6. Centrifuge the cells at 1380 × *g* for 5 min.
7. Remove supernatant, and wash cells with complete kidney culture medium.
8. Filter cells using 40 μm filters to remove any cell clumps (*see* **Note 12**).
9. Centrifuge the cells at 1380 × *g* for 5 min.
10. Remove supernatant, and to prepare cells for further culture, follow one of the protocols described in Subheading 3.6.

### 3.2 Separation of the Ureteric Bud from the Metanephric Mesenchyme

Once you have dissected out the embryonic kidneys, you can separate the UB from the MM.

1. Treat embryonic kidneys with trypsin/pancreatin solution in PBS−/− for 30–40 s at room temperature. It is important to

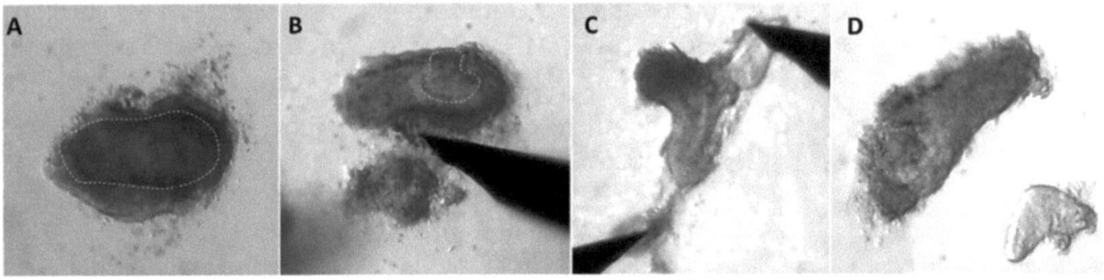


**Fig. 1** Separation of the UB from the MM. (**a**) Dissected out embryonic kidney. The dashed line surrounds the MM (everything outside the line is extra tissue to be removed). (**b**) After removal of the extra tissue surrounding the MM, the UB is visible (dashed line). (**c**) Removal of the UB from the MM. (**d**) Separated MM and UB

thaw a tube with trypsin/pancreatin solution just before application to embryonic kidneys.

2. Stop enzymatic reaction by removing the kidneys from trypsin/pancreatin solution into complete kidney culture medium (*see* Subheading 2.2 and **Note 2**). Try to transfer as little as possible of the trypsin/pancreatin solution.
3. Using dissecting microscope and needles (G27), gently remove all tissues surrounding the MM (*see* Fig. 1a, b), leaving only small amount to enable holding the kidney in place (docking place for needles). Use one needle to keep the kidney in place by poking into the extra tissue (to be removed later), and use another needle to remove the UB from the MM (*see* **Note 13**, Fig. 1c). Once the UB is out, remove that extra piece of tissue that you used to hold the kidney in place (*see* Fig. 1d).

### *3.3 Dissociation of the Metanephric Mesenchyme*

Once you have clean MM, you can dissociate it to single-cell suspension.

1. Transfer MM to low binding Eppendorf tube (*see* **Note 14**).
2. Treat MM with 320 μL of dissociation solution (*see* Subheading 2.3 and **Note 8**).
3. Place Eppendorf with MM into incubator (37 °C) for about 20–30 min.
4. Pipette vigorously every 7–10 min (*see* **Note 11**).
5. Once you can identify single cells down the microscope, stop the enzymatic reaction using 1 mL of complete kidney culture medium (*see* Subheading 2.3 and **Note 2**).
6. Pipette well.
7. Centrifuge at 1380 × *g* for 5 min. Remember to correctly balance the centrifuge. At the end of this step, the pellet of cells should be visible (*see* **Note 15**).
8. Carefully remove supernatant, and wash cells again with complete kidney culture medium.

9. Centrifuge at 1380 × *g* for 5 min.
10. Remove the supernatant and to prepare cells for experiment, follow one of the protocols described in Subheading 3.6.

### 3.4 Dissociation of Ureteric Bud

Once you have clean UB, you can dissociate it to single cell suspension.

1. Collect UBs into Eppendorf tube in small amount of medium, allow them to settle down, and then remove excess of medium from the tube being careful not to remove the UBs (*see* **Note 16**).
2. Add 1 mL of 1× trypsin/EDTA solution, mix UBs, and place them into water bath at 37 °C.
3. Incubate at 37 °C water bath for 10 min (*see* **Note 10**).
4. Pipette vigorously every 3–5 min (*see* **Note 11**).
5. Filter cells using 40 μm filters to remove any cell clumps (*see* **Note 12**).
6. The UB cells are ready to be used as inducer. Follow one of the protocols described in Subheading 3.6.

### 3.5 Dissection of Embryonic Spinal Cord

The embryonic spinal cord is one of the most potent nephrogenesis inducers [13].

1. Place cranial part of the embryo (head and upper body) on a glass dish filled with PBS−/−, and visualize under the microscope.
2. Use one needle to keep the embryo steady and the other one to separate the spinal cord from the rest of the embryo, starting right behind the brain stem (*see* **Note 17**, Fig. 2a).
3. Orientate the spinal cord placing the dorsal part to the dish in order to better visualize the spinal cord and somites. The spinal cord is a more dense tissue than somites, and it is therefore darker than somites surrounding it (*see* Fig. 2b).
4. Identify the border between the spinal cord and somites on the right-hand site, and cut along using one needle (*see* Fig. 2b (arrow) and c). Repeat the same cut on the other site of the spinal cord.
5. Cut the spinal cord into smaller pieces, and use them for tubulogenesis induction as described in Subheading 3.6.

### 3.6 Tubulogenesis Induction

It is well known that the MM undergoes apoptosis when the UB is not present and vice versa [2]. Therefore, to induce the tubulogenesis in the MM, it is necessary to provide the UB (whole or dissociated) or other tissue able to induce tubulogenesis, such as the embryonic spinal cord; or by activating the Wnt/β-catenin pathway chemically. Below, detailed protocol for each option is described.

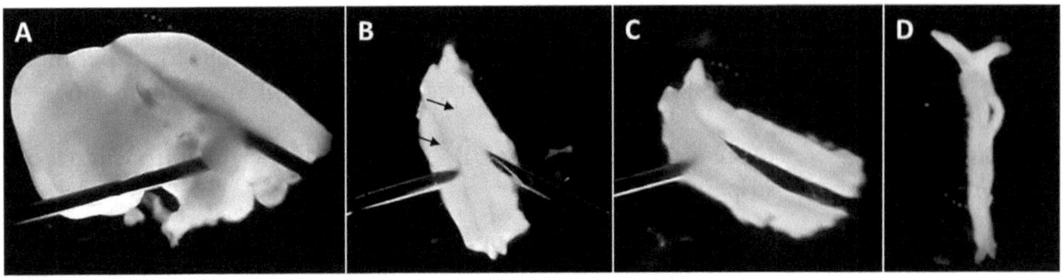


**Fig. 2** Dissection of the embryonic spinal cord. (**a**) The cranial part of the embryo position, to dissect out the spinal cord together with surrounding somites. (**b**) The spinal cord is located in between the somites. Border of these two tissues is visible and labeled with arrows. (**c**) Cut along the border of the spinal cord and somites. (**d**) Cleaned spinal cord

#### 3.6.1 With the Ureteric Bud

The induction of tubulogenesis in the MM with UB may be performed by applying few alternative routes (Fig. 3a–c).

##### Dissociation of Whole Embryonic Kidneys

1. Once you have dissociated embryonic kidney rudiments (*see* Subheading 3.1), resuspend the cells in 1 mL of complete kidney culture medium (*see* **Note 2**), and add ROCK inhibitor (Y27632) to achieve final concentration of 5 μM (*see* **Note 18**).
2. Count cells. Use between 100,000 and 50,000 cells to generate kidney pellets (*see* **Note 19**).
3. Transfer appropriate and equal amounts of medium with cells into new Eppendorf tubes.
4. If you want to add different cell types (exogenous cells) into this kidney culture, add them now (*see* **Note 20**).
5. Centrifuge the cells at 1380 × *g* for 5 min (*see* **Note 21**).
6. Incubate pellets at 37 °C and 5% $CO_2$ overnight.
7. Next day, use 1 mL pipette to collect the pellet from the Eppendorf tube (*see* **Note 22**), and transfer it into the upper compartment of the Transwell filter dish (*see* **Note 23**).
8. Transfer the plate with pellets into incubator (37 °C and 5% $CO_2$).

##### Induction of MM with Whole UB

Following separation of MM from UB, follow the protocol to dissociate the MM into single-cell suspension (*see* Subheading 3.3).

1. Transfer the UB into 3.5 cm culture dish.
2. Add 1 mL of complete kidney culture medium supplemented with GDNF (100 ng/mL), and incubate at 37 °C for 30 min.
3. In the meantime, take dissociated MM, and resuspend it in a complete kidney culture medium supplemented with BMP7 (50 ng/mL) and FGF2 (100 ng/mL) (*see* **Note 24**), and

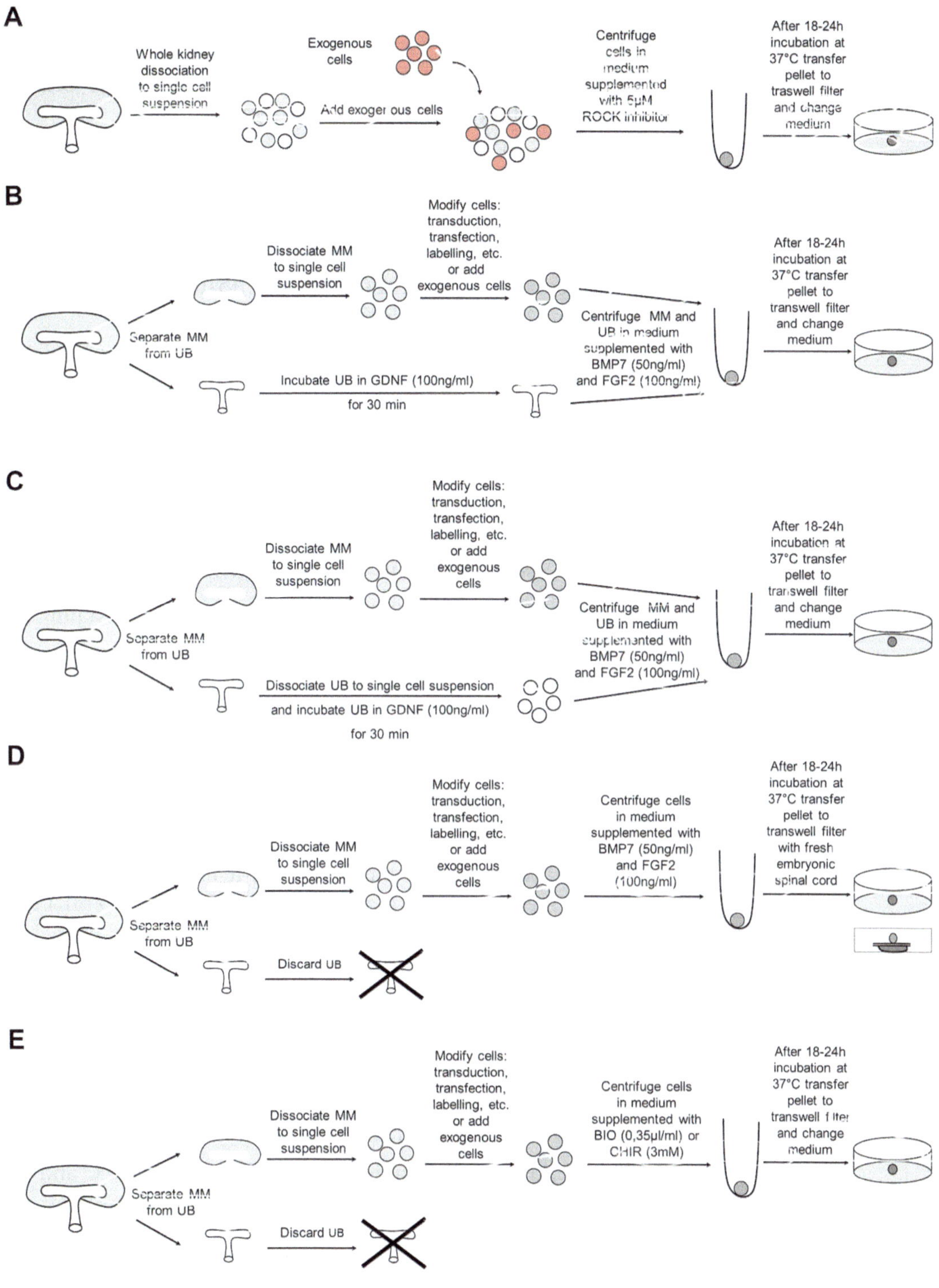


**Fig. 3** Schematic illustration of possible ways to induce tubulogenesis in metanephric mesenchyme. (**a**) Dissociation of whole embryonic kidneys—induction with dissociated UB, used to test nephrogenic potential of various exogenous cells (stem cells and others); (**b**) induction of MM with whole UB, enable modification of MM cells; (**c**) induction of MM with dissociated UB, enable modification of MM cells; (**d**) induction of MM with embryonic spinal cord, enable modification of MM; (**e**) induction of MM with chemicals, enable modification of MM (from [16]—modified)

count the cells. Use between 100,000 and 50,000 cells to generate kidney pellets (*see* **Note 19**).

4. Following 30 min incubation in GDNF, transfer two UBs into each Eppendorf with 50,000 cells of MM and mix well (*see* **Note 25**).
5. Centrifuge the cells at 1380 × *g* for 20 min.
6. Place centrifuged Eppendorfs into incubator at 37 °C and 5% $CO_2$ for 18–24 h (*see* **Note 22**).
7. The next day, transfer pellets into Transwell filters (*see* **Notes 22**, **23**, **26**, and **27**).
8. Transfer the plate with pellets into incubator (37 °C and 5% $CO_2$).

#### Induction of MM with Dissociated UB

Following separation of MM from UB, follow the protocol to dissociate the MM into single-cell suspension (*see* Subheading 3.3) and to dissociate UB into single cells (*see* Subheading 3.4).

1. Centrifuge the dissociated UB cells at 1380 × *g* for 3 min. To collect it at the bottom of the tube.
2. Gently remove supernatant, being careful not to remove the UB cells. Follow protocol in Subheading 3.4.
3. Add 500 μL of complete kidney culture medium supplemented with GDNF (100 ng/mL), mix well to resuspend the cells, incubate at 37 °C for 30 min, and count the cells. Use 1 UB cell for each 10 cells of the MM.
4. Take dissociated MM, and resuspend it in a complete kidney culture medium supplemented with BMP7 (50 ng/mL) and FGF2 (100 ng/mL) (*see* **Note 24**), and count the cells. Use between 100,000 and 50,000 cells to generate kidney pellets (*see* **Note 19**).
5. Transfer appropriate amount of MM cells into Eppendorf.
6. Transfer appropriate amount of UB cells into Eppendorf with the MM cells, keeping the ratio of 1:10. Mix well.
7. Centrifuge the cells at 1380 × *g* for 20 min.
8. Place centrifuged Eppendorfs into incubator at 37 °C and 5% $CO_2$ for 18–24 h (*see* **Note 22**).
9. The next day, transfer pellet onto Transwell filters (*see* **Notes 22, 23, 26**, and **27**).
10. Transfer the plate with pellets into incubator (37 °C and 5% $CO_2$).

#### *3.6.2 With the Embryonic Spinal Cord*

There are other tissues than UB known to induce tubulogenesis [14], and the embryonic spinal cord is one of the most potent inducers (Fig. 3d).

1. Following separation of MM from UB, follow the protocol to dissociate the MM into single-cell suspension (*see* Subheading 3.3).
2. Take dissociated MM, and resuspend it in a complete kidney culture medium supplemented with BMP7 (50 ng/mL) and FGF2 (100 ng/mL) (*see* **Note 24**), and count the cells. Use between 100,000 and 50,000 cells to generate kidney pellets (*see* **Note 19**).
3. Transfer appropriate and equal amounts of medium with cells into new Eppendorf tubes.
4. Centrifuge the cells at 1380 × *g* for 20 min.
5. Place centrifuged Eppendorfs into incubator at 37 °C and 5% $CO_2$ for 18–24 h (*see* **Note 22**).
6. The next day, dissect embryonic spinal cord (E11.5) following directions in Subheading 3.5. Once you have smaller pieces of spinal cord, place them into Transwell filters (*see* **Note 23**).
7. Place Millipore 0.1–1.0 μm filter on top of the spinal cord (*see* **Note 28**).
8. Transfer pellet from Eppendorf tube onto fresh Millipore 0.1–1.0 μm filter (*see* **Note 29**), and then transfer it to the Transwell filter with spinal cord and Millipore filter. Position the MM directly (or as close as possible) on top of the spinal cord.
9. Transfer the plate with pellets into incubator (37 °C and 5% $CO_2$).

*3.6.3 With Chemicals (BIO, LiCl)*

It is also possible to replace any nephrogenesis-inducing tissue with chemicals that inhibit GSK-3. This leads to cytoplasmic accumulation of β-catenin, which results in activation of target genes of the Wnt/β-catenin pathway (Fig. 3e).

1. Following separation of MM from UB, follow the protocol to dissociate the MM into single-cell suspension (*see* Subheading 3.3).
2. Take dissociated MM, and resuspend it in a complete kidney culture medium supplemented with BMP7 (50 ng/mL) and FGF2 (100 ng/mL) (*see* **Note 24**), and count the cells. Use between 100,000 and 50,000 cells to generate kidney pellets (*see* **Note 19**).
3. Transfer appropriate number of cells into Eppendorfs.
4. Centrifuge the cells at 1380 × *g* for 20 min.
5. Following centrifugation, gently remove as much of the medium as possible, and induce the pellets by adding complete kidney culture medium supplemented with either BIO (5–10 μM) [11] or LiCl (15 mM) [9]. Add it gently to not disrupt pellets (*see* **Note 30**).

6. Place centrifuged Eppendorfs into incubator at 37 °C and 5% $CO_2$ for 18–24 h (*see* **Notes 22** and **31**).
7. The next day, transfer pellets onto Transwell filters (*see* **Notes 23**, **26**, and **27**).
8. Transfer the plate with pellets into incubator (37 °C and 5% $CO_2$).

## 4 Notes

1. To dissect embryonic kidneys, follow the protocols described in Chapters 2 and 3 in this MiMB book.
2. To prepare 100 mL of complete kidney culture medium, use 10 mL of 100% fetal calf serum, 1 mL of penicillin/streptomycin solution (bought ready to use from, e.g., Sigma), and fill up to 100 mL with DMEM high glucose (4.5 mg/mL).
3. Filtering the medium prevents it from fungal contamination and extends its use.
4. To prepare pancreatin solution, measure 0.25 g of pancreatin (Sigma), 0.17 g of NaCl, and 20 mL of $dH_2O$. Mix all components using magnetic stirrer for 3–4 h at RT. Then transfer to 4 °C O/N. Next day, centrifuge the solution for 10 min at 2700 × *g* at RT, and filter the solution using 0.22 μm Millipore syringe filters. Make 2 mL aliquots and store them at −20 °C until needed.
5. To prepare Tyrode's solution, measure 8 g of NaCl, 0.2 g of KCl, 0.05 g of $NaH_2PO_4$ (MW 137.99), 1 g of glucose ($H_2O$ free, D (+) glucose, MW 180.16), 1 g of $NaHCO_3$, 1 L $dH_2O$.- Dissolve all components, and filter solution using 0.22 μm Millipore syringe filters. Store at RT until needed.
6. The pH should be in range between 7.2 and 7.8.
7. To prepare 100 mL of 0.5% BSA in PBS (−/−), measure 0.5 g of BSA (Sigma) and add 100 mL of PBS calcium/magnesium free (−/−).
8. To prepare collagenase IV solution, to the original tube containing 100 mg of collagenase 4 (Worthington Biochemical Corporation), add 4 mL of Earle's Balanced Salt Solution, mix well, and make small aliquots of ready collagenase 4 solution.
9. If older kidney rudiments are used (e.g., E13.5 [15]), cut them with a scalpel into small pieces (~1 mm × 1 mm) before placing them into the trypsin/EDTA solution. This will improve dissociation.

10. It is important to use water bath as it is better in transferring the heat. It is also important not to prolong this incubation step as trypsin may harm the cells.
11. The pipetting should separate cell treated with trypsin/collagenase from the outer edge of the pieces; however, it should not destroy cells that are already single in the solution.
12. Failing to filter the cell suspension may result in clumps of cells remaining in the suspension and generating false-positive results. This may happen, because ureteric bud is an epithelial structure and is difficult to dissociate into single-cell suspension, and given that it is also a potent inducer of the metanephric mesenchyme, the induction may take place sooner, in comparison to tissues completely devoid of ureteric bud cell clumps.
13. When separating UB from the MM with needles, try to use one needle to keep the MM in place, and use the other one to remove the UB from it. If the trypsin/pancreatin solution was fresh and the kidneys were treated long enough, the UB should come out of the MM intact. If you have troubles to separate the UB from the MM in one piece, you should wash the remaining pieces of kidneys with PBS−/− and treat again with fresh trypsin/pancreatin solution.
14. Using low binding Eppendorf tubes enables easy transfer of the cellular pellet, especially in cases when the pellet is very small.
15. If you place all your tubes in the centrifuge in the same direction, e.g., the tube opening toward the inside while the top holder into outer site of the centrifuge, then you should be able to identify the site of probable cell pellet.
16. The UB is quite small tissue sample, and one litter may not give you enough cells to perform induction experiments. We usually use UB from 2 or 3 litters of mice.
17. This will separate the spinal cord from the rest of the body, but it will still be connected to somites that surround it. They need to be cleaned away from the spinal cord to not interfere with induction.
18. ROCK inhibitor is required only during the first 24 h of culture to support epithelial cells after dissociation, but it should be removed from culture after that time elapses as in the later stages of kidney culture it is blocking the nephron development [7].
19. It is possible to generate pellets from smaller number of cells (e.g., 15,000), but for unexperienced person, these are very difficult to transfer from Eppendorf tube into filter.
20. Prepare the cells in a way that they would be in single-cell suspension (e.g., by trypsinization), also count the number of

cells you want to use, and then transfer appropriate amount into tubes with dissociated kidney cells.

21. At the end of this step, you should be able to see the cell pellet that will have a shape of triangle with wide base at the bottom of the Eppendorf and a pick at the top.
22. During the overnight incubation in the Eppendorf tubes, the pellets change shape from flat to more round/spherical; this is easily detachable from the centrifuge tube.
23. Use 1 μm hole size to enable easy medium flow from top to bottom compartment of the Transwell filter dish. We use 12-well dishes. Fill the bottom compartment with complete kidney culture medium until small amount of medium is present also in the top compartment.
24. This is also a good time if one would like to manipulate the MM cells by transduction, transfection, or simply labeling the cells. In these conditions MM cells are viable and competent to respond to inductive signals for about 24 h.
25. We use two UBs for 50,000 cells; therefore, if you use more MM cells to generate the pellet, use accordingly more UBs to induce it.
26. Collect the pellet from the medium using 1 mL pipette trying to collect minimal amount of medium, and transfer it to the upper compartment of the Transwell filter dish.
27. Following that incubation time, it is recommended to change medium; we do it by transferring the pellets into new dishes (Transwell filter dishes) with fresh complete kidney culture medium [7].
28. This step may be omitted, but it will make it easier to separate the spinal cord from the induced MM for immunostaining and imaging purposes. At the end, the layers you will have will include (a) Transwell filter membrane, (b) embryonic spinal cord, (c) Millipore filter, (d) Millipore filter, and (e) MM pellet.
29. When you transfer the pellet onto Millipore filter, it is good to use open forceps to hold the filter on the dish (in two places), during transfer. This is to avoid situation in which the filter would move and get stuck to the pipette due to electrostatic forces and therefore unable the pellet transfer.
30. If you happen to disrupt the pellet when adding the induction medium, just centrifuge again at 1380 × *g* for 20 min.
31. We do not recommend exceeding 24 h mark of incubation with BIO.

## Acknowledgments

I would like to thank Sanna Junttila and Paula Haipus for producing the movies from which Figs. 1 and 2 were made.

### References

1. Saxen L (1971) Inductive interactions in kidney development. Symp Soc Exp Biol 25:207–221
2. Grobstein C (1953) Inductive epitheliomesenchymal interaction in cultured organ rudiments of the mouse. Science 118(3054):52–55
3. Auerbach R, Grobstein C (1958) Inductive interaction of embryonic tissues after dissociation and reaggregation. Exp Cell Res 15(2):384–397
4. Trowell OA (1954) A modified technique for organ culture in vitro. Exp Cell Res 6(1):246–248
5. Saxen L, Koskimies O, Lahti A, Miettinen H, Rapola J, Wartiovaara J (1968) Differentiation of kidney mesenchyme in an experimental model system. Adv Morphog 7:251–293
6. Unsworth B, Grobstein C (1970) Induction of kidney tubules in mouse metanephrogenic mesenchyme by various embryonic mesenchymal tissues. Dev Biol 21(4):547–556
7. Unbekandt M, Davies JA (2010) Dissociation of embryonic kidneys followed by reaggregation allows the formation of renal tissues. Kidney Int 77(5):407–416
8. Kispert A, Vainio S, McMahon AP (1998) Wnt-4 is a mesenchymal signal for epithelial transformation of metanephric mesenchyme in the developing kidney. Development 125(21):4225–4234
9. Davies JA, Garrod DR (1995) Induction of early stages of kidney tubule differentiation by lithium ions. Dev Biol 167(1):50–60
10. Halt K, Vainio S (2012) In vitro induction of nephrogenesis in mouse metanephric mesenchyme with lithium introduction and ureteric bud recombination. Methods Mol Biol 886:23–30
11. Kuure S, Popsueva A, Jakobson M, Sainio K, Sariola H (2007) Glycogen synthase kinase-3 inactivation and stabilization of beta-catenin induce nephron differentiation in isolated mouse and rat kidney mesenchymes. J Am Soc Nephrol 18(4):1130–1139
12. Junttila S, Saarela U, Halt K, Manninen A, Parssinen H, Lecca MR, Brandli AW, Sims-Lucas S, Skovorodkin I, Vainio SJ (2015) Functional genetic targeting of embryonic kidney progenitor cells ex vivo. J Am Soc Nephrol 26(5):1126–1137
13. Lombard MN, Grobstein C (1969) Activity in various embryonic and postembryonic sources for induction of kidney tubules. Dev Biol 19(1):41–51
14. Rak-Raszewska A, Hauser PV, Vainio S (2015) Organ in vitro culture: what have we learned about early kidney development? Stem Cells Int 2015:959807
15. Rak-Raszewska A, Wilm B, Edgar D, Kenny S, Woolf AS, Murray P (2012) Development of embryonic stem cells in recombinant kidneys. Organogenesis 8(4):125–136
16. Rak-Raszewska A, Vainio S (2016) Nephrogenesis in organoids to develop novel drugs and progenitor cell based therapies. Eur J Pharmacol 790:3–11

# Part II

## Embryonic Kidney Progenitor Cell and 3D Kidney Model Cell Like Based Cultures

# Chapter 5

# Dissociation of Tissues for Single-Cell Analysis

Andrew S. Potter and S. Steven Potter


## Abstract

It is now a routine to carry out single-cell RNA-Seq to define the gene expression patterns of thousands of cells, thereby revolutionizing many areas of research. Projects are underway to use these techniques to create an atlas of the expressed genes in all cell types of the human body. Here we describe cold-active protease methods for single-cell dissociation of organs and tissues that better preserve the in vivo gene expression patterns.


**Key words** Organ dissociation, Single-cell suspensions, Psychrophilic proteases, Reduced gene expression artifacts, Immediate early response genes

## 1 Introduction

The breakdown of an organ or tissue to a single-cell suspension, to allow single-cell analysis, can be a lengthy and challenging process. Typically, this includes the use of proteases to help digest the attachments between cells. Examples of proteases often used include trypsin [1, 2], papain [3], pronase, TrypLE, elastase [4], collagenase [5], neutral protease (dispase) [6, 7], hyaluronidase [8], and liberase, a commercial cocktail. All of these enzymes, however, are used at 37 °C, a temperature where the enzymes of the mammalian cells being dissociated are also optimally active. It can be expected, therefore, that the dissociation procedure, which places the cells in a very foreign environment and has historically often required protease digestion periods of 30 min, 60 min, or sometimes more, can result in artifact changes in gene expression for the cells of interest. Indeed, it has been shown that such dissociation methods can cause gene expression changes of over a thousandfold for some genes.

We reasoned that if it were possible to carry out the tissue dissociation at greatly reduced temperatures, or even on ice, then this would result in fewer artifact changes in gene expression patterns [9]. Just as there are thermophilic organisms adapted to live at

Seppo Vainio (ed.), *Kidney Organogenesis: Methods and Protocols*, Methods in Molecular Biology, vol. 1926,
https://doi.org/10.1007/978-1-4939-9021-4_5, 

elevated temperatures, such as thermal vents and hot springs, there are also psychrophilic microorganisms adapted to survive in very cold environments, such as mountaintops and glaciers. These microorganisms have enzymes that maintain relatively high activity even in the cold. Here we describe dissociation methods we have developed that use protease from *Bacillus licheniformis*, which is a soil bacteria that has been isolated growing on Himalayan glaciers. The dissociation methods using this enzyme can be carried out at low temperatures, and even on ice, thereby greatly reducing artifact gene expression changes.

Different tissues require different dissociation protocols. There is no one-size-fits-all solution. In general, embryonic tissues dissociate easily, while adult tissues are more challenging. Diseased and fibrotic tissues can be especially difficult to dissociate. Below we present three dissociation protocols using cold-active protease that can be carried out at near ice temperatures.

## 2 Materials

It is best to maintain tissue on ice or at near ice temperatures from start to finish. When possible remove tissue of interest rapidly from subject and cool in ice-cold DPBS. Gene expression changes appear to be minimal at ice-cold temperature. Tissue can be shipped on ice overnight with few changes in gene expression patterns. After a single-cell preparation is made, the cells can be frozen as per tissue culture cells, in 10% DMSO, 50% fetal bovine serum, and 40% DMEM, cooling slowly in a Styrofoam container placed at −80 °C overnight and then transferred to liquid nitrogen the next day. Cells can then be thawed rapidly in a 37 °C water bath, and for many cell types, viability post freeze/thaw is over 90%.

### 2.1 Stock Solutions

1. Protease stock solution. Weigh 100 mg of *Bacillus licheniformis* protease (Sigma P5380) and add to 1 mL of DPBS (Ca- and Mg-free), mix to dissolve, and rapid freeze 100 μL aliquots by placing tubes in powdered dry ice. Store at −80 °C. Enzyme retains activity for at least 6 months (*see* **Note 1**).
2. Bovine serum albumin stock solution. Make 10 mg/mL in ultrapure water and store at −80 °C.
3. DNase stock solution with 250 U/10 μL in PBS stored at −80 °C.
4. Red blood cell lysis solution can be prepared with 0.8 g $NH_4Cl$ (ammonium chloride), 84 mg of $NaHCO_3$ (sodium bicarbonate), and 37 mg of EDTA (disodium), QS to 100 mL with ultrapure water, or can be purchased from Sigma (R7757).
5. DPBS/BSA. This is DPBS with 0.01% added BSA (bovine serum albumin). Make fresh.

6. 0.5 m EDTA.
7. Trypan blue, 0.4%.
8. DMEM for freezing can be purchased from Gibco.

### 2.2 Equipment and Supplies

1. Eppendorf refrigerated thermomixer C or R (5382000015/Z605271).
2. Miltenyi gentleMACS Dissociator (*see* **Note 2**).
3. Refrigerated centrifuge.
4. Cell strainers, 30, 40, 70, and 100 μm, can be purchased from Falcon and Miltenyi.

## 3 Methods

### 3.1 Newborn Mouse Kidneys

1. Harvest kidneys rapidly and cool briefly in ice-cold DPBS.
2. Place kidneys on a plastic petri dish that is resting on ice.
3. Take a fresh razor blade and mince the tissue thoroughly. In our experience autoclaving of the razor will dull the blade, so use without autoclaving. Adding a small cutting motion to the mincing chopping of the razor can help cut through connective tissue. Thorough mincing helps to increase the surface area that will be available to the enzyme and reduces the size of the tissue blocks to be digested.
4. Weigh out 25 mg of minced tissue and add to a conical 1.5 mL conical snap cap plastic tube on ice.
5. Make 1× enzyme digestion buffer by adding 100 μL of *B. licheniformis* stock to 895 μL of DPBS and 5 μL of 0.5 M EDTA.
6. Add 1× enzyme digestion solution to kidney, mince, and quickly shake thoroughly to mix and resuspend mince.
7. Incubate in ice water bath.
8. Triturate using Pipetman with 1 mL pipet tip for 15 s every 2 min (*see* **Note 3**).
9. After three rounds of trituration, a total of about 7 min on ice, transfer to a Miltenyi C-tube on ice.
10. Using the Miltenyi GentleMACS, in the cold room, run the brain_03 program twice.
11. Return sample to 1.5 mL conical tube on ice, and incubate an additional approximate 7 min with 15 s trituration every 2 min interval (*see* **Note 4**).
12. Filter cells through a 30 μm strainer, rinse strainer with 10 mL of ice-cold PBS/BSA, and transfer cells to 14 mL conical tube.
13. Centrifuge cells 600 *g* for 5 min at 4 °C. Discard supernatant, and resuspend pelleted cells in 1 mL of ice-cold PBS/BSA.

14. Determine cell concentration using a hemocytometer, and adjust to desired cell concentration for subsequent single-cell RNA-Seq by appropriate dilution with PBS/BSA or by again pelleting and resuspending.
15. If necessary the red blood cells can be removed by pelleting the cells (600 *g* 5 min 4 °C), resuspending the pellet in 1 mL of RBC lysis buffer, triturating 30× using a 1 mL Pipetman, and incubating on ice for 2 min. Then add 12 mL of ice-cold DPBS/BSA and re-pellet cells.

### 3.2 Adult Mouse Lung

1. Harvest and mince tissue on ice until fine paste as described for newborn kidney.
2. Make two tubes of 1 mL each 1× enzyme mix by adding to each tube 150 μL of *B. licheniformis* enzyme 100 mg/mL stock, 1 μL 0.5 M EDTA, and 894 μL of DPBS (Ca- and Mg-free).
3. Place 25 mg tissue in 1 mL enzyme mix in 1.5 mL conical snap cap tube on ice.
4. Shake tube briefly to mix contents once every 30 s. Begin triturating at 2 min. Triturate 10× (ten strokes with 1 mL Pipetman set at 700 μL) every minute.
5. After 5 min pipet tissue plus enzyme mix into petri dish on ice. Mince again 1–2 min to further break up tissue.
6. Pipet tissue plus enzyme mix back into 1.5 mL tube. Rinse petri dish with 0.5 mL PBS/0.5 mM EDTA. Pipet into tube with tissue and enzyme (new concentration: 10 mg/mL *B. licheniformis* protease).
7. Continue triturating on ice, ten strokes each minute, using 1 mL Pipetman, for 10 min.
8. At 10 min centrifuge at 15 *g* for 10 s at 4 °C to pellet undigested chunks of tissue. Pipet ~80% of supernatant onto 40 μm strainer on sterile 50 mL conical tube.
9. Rinse strainer with 6 mL ice-cold PBS/BSA. Keep filtered cells on ice. Leave filter on 50 mL conical tube for later steps.
10. Add additional 1 mL of 15 mg/mL enzyme mix to residual clumps of tissue in enzyme in the 1.5 mL tube.
11. Continue triturating on ice for 10 min (20 min total digestion time). Pipet entire volume to original 40 μM filter, and rinse with 7 mL ice-cold DPBS/BSA. Transfer flow-through to 15 mL conical.
12. Centrifuge 650 *g* 4 °C for 5 min. Remove all but 100 μL of supernatant, being careful not to disturb pellet.
13. Add 900 μL RBC lysis buffer to pellet. Triturate 30× using 1 mL pipet and incubate for 2 min on ice.

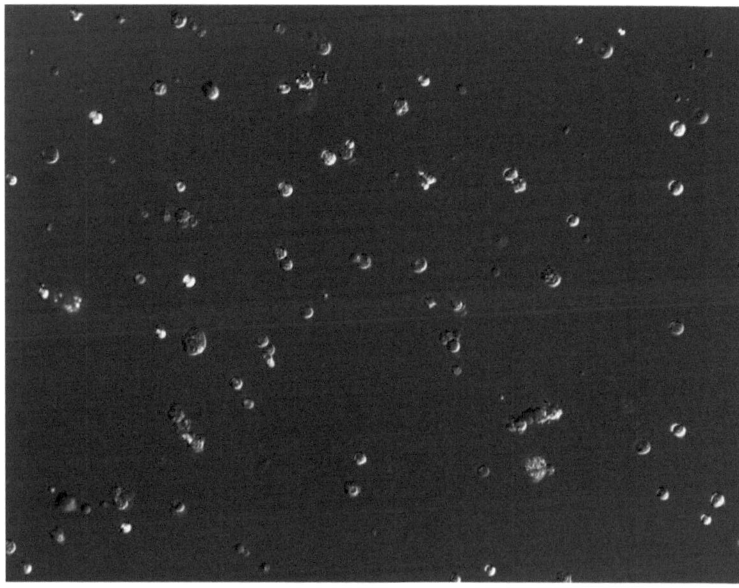

**Fig. 1** Dissociation of adult mouse lung. While most cells are singlets, there remain some doublets and clumps of multiple cells. The variety of cell sizes reflects the diversity of cell types present in the mouse adult lung. Over 90% of cells are viable as determined by trypan blue staining (not shown)

14. Add 12 mL PBS/BSA to 50 m conical tube.
15. Centrifuge 650 *g* 4 °C for 5 min. Remove supernatant and resuspend pelleted cells in 700 μL PBS/BSA. Examine using hemocytometer with added equal volume of trypan blue stock solution to monitor single-cell dissociation, cell concentration, and viability (Fig. 1) (*see* **Note 5**).

### 3.3 Adult Mouse Kidney

1. Quickly remove adult kidney; cool rapidly ice-cold DPBS.
2. Remove fatty tissue and kidney capsule in ice-cold PBS.
3. Finely mince on petri dish, on ice (~2 min).
4. Weigh out 25 mg tissue and add to 1 mL enzyme mix (100 μL *B. licheniformis* protease stock, 1 μL 0.5 M EDTA, 894 μL DPBS, 5 μL DNase stock).
5. Shake briefly every 30 s to mix tissue and enzyme.
6. After 2 min triturate gently with 20× strokes using 1 mL Pipetman set to 700 μL.
7. For an additional 8 min, alternate shaking every other minute to resuspend cell clumps and trituration every other minute with 10× strokes using 1 mL Pipetman set to 700 μL.
8. After total of 11.5 min, spin briefly, for 10 s, 50 *g*, 4 °C to pellet cell clumps and remove ~80% of supernatant and examine with microscope to insure presence of single-cell suspension.

9. Pass supernatant through 30 μm cell strainer into 50 mL conical tube on ice, and rinse strainer with 8 mL of ice-cold DPBS/BSA.
10. Add 1 mL of enzyme mix (same as before) to pellet, triturate 10× strokes with 1 mL Pipetman.
11. Place tube in refrigerated thermomixer in cold room with final temperature at 4 °C set to 1200 RPM.
12. Every 3 min passage tissue 18× through an 18 gauge needle with a 1 mL syringe, for a total of three times.
13. Once again, spin at 50 *g*, 4 °C, 10 s to spin down cell clumps.
14. Remove ~80% of supernatant pass through 30 μm strainer. Rinse with 8 mL ice-cold DPBS/BSA into the same 50 mL conical as first round digestion cells.
15. Add additional 1 mL enzyme mix to residual tissue.
16. Continue digesting in thermomixer at 1400 RPM in at 4 °C.
17. Every 3 min passage 8× with 18 gauge needle with 1 mL syringe (3× total).
18. Triturate 10× and apply to total volume to the same 30 μm strainer used in previous steps. Rinse filter with 8 mL ice-cold DPBS/BSA.
19. Transfer combined strainer flow-through to two 15 mL conicals. Spin 500 *g*, 6 min, 4 °C.
20. Remove supernatant; resuspend combined pellets in 100 μL total volume ice-cold DPBS/BSA, and add 900 μL RBC lysis buffer. Triturate 20×. Let sit 2 min on ice.
21. Add additional 9 mL DPBS/BSA. Spin 500 *g*, 4 °C for 5 min.
22. Resuspend pellet in 1 mL DPBS/BSA. Examine using hemocytometer with added equal volume of trypan blue stock solution to monitor single-cell dissociation, concentration, and viability (*see* **Notes 6** and 7).

## 4 Notes

1. We have tested several cold active psychrophilic proteases, including Atlantic cod trypsin, and proteases from *Bacillus* sp. and *Bacillus amyloliquefaciens*. In our hands, the *B. licheniformis* protease worked the best, although *Bacillus* sp. protease was not far behind and is much less expensive. We've also found that collagenases A and D from *Clostridium histolyticum* (Roche), coupled with trypsin inhibitor (Roche), can be particularly effective at ice-cold dissociation of very difficult tissues.

2. The Eppendorf thermomixer and Miltenyi gentleMACS Dissociator provide useful tools to augment the enzymatic dissociation, but if these are not available, then simple trituration using a Pipetman can substitute.
3. It might be necessary to cut the Pipetman 1 mL tip to make the hole bigger to prevent clogging, depending on how thorough the mincing was. If so then return to an uncut tip when possible without clogging to maximize dissociation.
4. During this incubation monitor success of dissociation with a microscope. This is critically important. Adjust accordingly. If cells are almost entirely single cell after only a few minutes, then proceed to the next step. If the cells are not yet fully dissociated, then continue beyond the recommended time for dissociation.
5. Some cells will dissociate more easily than others, and some cells are more easily killed by the dissociation process than others. The net result is that the final single-cell suspension will not provide accurate representation of the cell-type proportionality present in the initial tissue. Don't expect it, and don't think the final cell type counts derived from the scRNA-Seq data will accurately define percent cell types in the tissue of interest.
6. We refer to this procedure as a layered approach. The tissue is subjected to repeated rounds of enzymatic digestion, with removal of released cells, followed by additional digestion of residual cell clumps to dissociate more cells. In this manner, the easy to dissociate cells are removed first and stored in diluted enzyme on ice to reduce the chances that they are killed by the subsequent harsher digestion conditions needed to remove the more difficult to dissociate cells. The cells are removed layer by layer. The goal is to preserve viability of the more easily removed cells while using ever more rigorous methods to separate out the most difficult to release cells, like podocytes and mesangial cells.
7. Where does one draw the line in quality assessment for continuation of single-cell analysis? This will vary for different tissues. If tissue is abundant and readily available for repeat experiments, then the acceptable quality metrics can be set high, for example, requiring that over 95% of cells are alive and actually single cell, not clumped. On the other hand, for rare patient samples, one might set the bar much lower, accepting for example 60% viability and a relatively large proportion of cell doublets and clumps. Cell doublets can confound the eventual data analysis for it can be very difficult to distinguish a doublet from an intermediate cell type. For most single-cell scRNA-Seq technologies, such as 10× Genomics and Drop-Seq, there will

be some doublets, with two cells in a drop, simply because of the Poisson distribution nature of the distribution of cells into drops. But many "single-cell" preparations actually include a large number of undissociated cells, which can, again, confound analysis. We therefore prefer to push the dissociation to give a higher percent of single cells, even though the cost is more cell death. Additional protocols can be found at protocols.io with search for Potter.

## Acknowledgments

This work was supported by NIH UO1 HL 122642 and UO1 DK107350 (SSP).

### References

1. Tuttle JB, Suszkiw JB, Ard M (1980) Long-term survival and development of dissociated parasympathetic neurons in culture. Brain Res 183 (1):161–180
2. Singh SK et al (2004) Identification of human brain tumour initiating cells. Nature 432 (7015):396–401
3. Panchision DM et al (2007) Optimized flow cytometric analysis of central nervous system tissue reveals novel functional relationships among cells expressing CD133, CD15, and CD24. Stem Cells 25(6):1560–1570
4. Barkauskas CE et al (2013) Type 2 alveolar cells are stem cells in adult lung. J Clin Invest 123 (7):3025–3036
5. Gentile P et al (2012) Concise review: adipose-derived stromal vascular fraction cells and platelet-rich plasma: basic and clinical implications for tissue engineering therapies in regenerative surgery. Stem Cells Transl Med 1 (3):230–236
6. Yan Y et al (2013) Implication of 14-3-3 epsilon and 14-3-3 theta/tau in proteasome inhibition-induced apoptosis of glioma cells. Cancer Sci 104(1):55–61
7. Volovitz I et al (2016) A non-aggressive, highly efficient, enzymatic method for dissociation of human brain-tumors and brain-tissues to viable single-cells. BMC Neurosci 17:30
8. Sawamura Y et al (1988) Isolation and in vitro growth of glioma-infiltrating lymphocytes, and an analysis of their surface phenotypes. J Neurosurg 69(5):745–750
9. Adam M, Potter AS, Potter SS (2017) Psychrophilic proteases dramatically reduce single-cell RNA-seq artifacts: a molecular atlas of kidney development. Development 144 (19):3625–3632

# Chapter 6

# Long-Term Culture of Nephron Progenitor Cells Ex Vivo

**Aaron C. Brown, Ashwani K. Gupta, and Leif Oxburgh**

## Abstract

Nephrons differentiate from the cap mesenchyme of the fetal kidney. Nephron progenitor cells that populate the cap mesenchyme efficiently balance self-renewal and epithelial differentiation to enable repeated rounds of nephron formation during development. Here we describe a method to isolate and propagate these cells from the embryonic mouse kidney. Using this method, nephron progenitor cells from a single litter of mice can be propagated to hundreds of millions of cells that express appropriate markers of the undifferentiated state and retain epithelial differentiation capacity in vitro.

**Key words** Kidney, Organogenesis, Primary cell culture, Stem cells

## 1 Introduction

Lineage tracing studies using genes expressed in the cap mesenchyme of the developing kidney have shown that this population gives rise to all epithelial cells of the nephron [1, 2]. Our group has been interested in studying growth factor signaling pathways that govern renewal of these nephron progenitor cells (NPCs) for several years, and we established a protocol for their propagation to test if our understanding of cell signaling was sufficiently comprehensive to reproduce the cap mesenchyme environment ex vivo [3–6]. Our studies showed that NPCs could indeed be propagated in their undifferentiated state while retaining their potential for epithelial differentiation. An application derived from our studies that has been of practical use to our group and others has been the NPC propagation protocol. Using this method we have been able to propagate cells up to at least 15 doublings in their undifferentiated state with retained capacity to differentiate (*see* **Note 1**). The protocol we provide here is in most aspects unchanged from our initial method, but we include our experiences with various reagents, including the basal medium (APEL), which is no longer commercially available (*see* **Note 2**). A summary of the procedure is shown in Fig. 1.

Seppo Vainio (ed.), *Kidney Organogenesis: Methods and Protocols*, Methods in Molecular Biology, vol. 1926, https://doi.org/10.1007/978-1-4939-9021-4_6, © Springer Science+Business Media, LLC, part of Springer Nature 2019

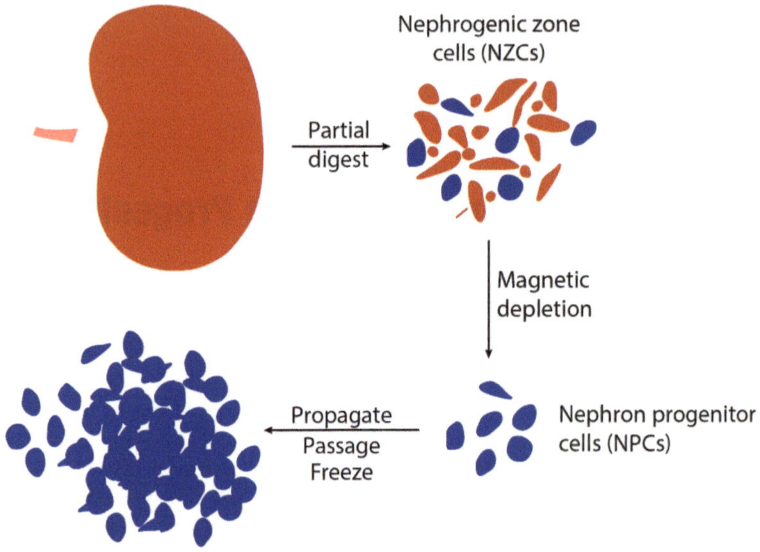


**Fig. 1** Schematic representation of the isolation and propagation procedure

## 2 Materials

### 2.1 Equipment

1. Stereomicroscope.
2. Watchmaker's forceps.
3. Nutator placed at room temperature (RT).
4. Nutator placed at 37 °C.
5. MidiMACS Separator (Miltenyi Biotec).
6. MACS MultiStand (Miltenyi Biotec).
7. Hemocytometer.

### 2.2 Embryo Dissection

1. Sterile 10 cm petri dishes.
2. Phosphate buffered saline (PBS) without calcium or magnesium.
3. Watchmaker's forceps, sterilized.
4. Sterile 3 mL plastic transfer pipettes.

### 2.3 Digestion of Kidneys

1. Collagenase A/pancreatin enzyme digest solution: Add 25 mg collagenase A to 10 mL PBS without calcium or magnesium. Place on a nutator at RT for 15 min. Add 100 mg porcine pancreatin and place on nutator at RT for 1.5 h. Filter sterilize (0.2 μm filter) before use to remove any undissolved particles. Batches of collagenase A/pancreatin enzyme digest solution can be made, filter sterilized, and frozen at −20 °C for up to 6 months. Incubation times vary depending on the strength of the enzyme cocktail, and the digest time should be empirically

determined and adjusted for each lot of solution by performing a timed digestion of E17.5 kidneys. Groups of six kidneys are digested for 10, 12, 14, and 16 minutes, and the dissociated cells are counted. The isolation of more than 1.75 million NZCs from six kidneys indicates over-digestion.

2. Round bottom 5 mL tubes.
3. Microfuge tubes.
4. 15 mL conical tubes.
5. 30 μm pre-separation filters (Miltenyi Biotec).
6. PBS-BSA: A solution of PBS (phosphate buffered saline) pH 7.2, supplemented with 0.5% BSA and 2 mM EDTA.
7. Hank's balanced salt solution (HBSS).
8. Fetal bovine serum (FBS).

#### 2.4 Magnetic Separation of NPCs (See Note 3)

1. Reagents for MACS immunomagnetic separation (Miltenyi Biotec): autoMACS running buffer, LS Columns.
2. Antibodies for MACS immunomagnetic separation (Miltenyi Biotec): anti-CD140-PE (130-102-502), anti-Ter119-PE (130-102-893), anti-CD326-PE (130-102-265), anti-CD105-PE (130-102-548), anti-PE microbeads (130-048-801).

#### 2.5 Cell Culture

1. Make nephron progenitor expansion medium (NPEM) according to the concentrations given in Table 1 in a laminar flow cabinet for cell culture, and thoroughly mix by inverting several times. The following reagents are used in the medium (*see* **Note 4**): APEL 2 Medium (Stemcell Technologies), Protein Free Hybridoma Medium (PFHM-II) (Thermo Fisher), FGF9 (R&D, 273-F9-025), BMP7 (R&D, 354-BP-010), BMP4 (R&D, 314-BP-010), IGF1 (R&D, 791-MG-050), IGF2 (R&D, 792-MG-050), CHIR99021 (Stemgent, 04-0004), LDN-193189 (Stemgent, 04-0074-02), Y27632 (Millipore, 688001), Heparin (Sigma, H3393).
2. 96-, 24-, or 6-well tissue culture plates.
3. Matrigel, growth factor reduced.
4. DMEM/F12 medium.
5. TrypLE™ dissociation solution (Life Technologies).
6. CryoStem Freezing Medium (Stemgent).
7. Cryovials.
8. 15 mL conical tubes.

**Table 1**
**Nephron progenitor expansion medium (NPEM)**

| Factor | Stock | Working |
|---|---|---|
| APEL 2 medium | 1× | 1× |
| PFHM-II | 1× | 0.05× |
| Pen Strep | 100× | 1× |
| FGF9 | 100 μg/mL | 200 ng/mL |
| Y27632 | 5 mM | 10 μM |
| IGF1 | 1 mg/mL | 20 ng/mL |
| IGF2 | 1 μg/mL | 2 ng/mL |
| BMP4 | 100 μg/mL | 30 ng/mL |
| BMP7 | 100 μg/mL | 30 ng/mL |
| LDN-193189 | 10 mM | 125 nM |
| CHIR99021 | 10 mM | 1.25 μM |
| Heparin | 10 mg/mL | 1 μg/mL |

## 3 Methods

### 3.1 Preparation of Reagents Prior to Isolation

1. Make collagenase A/pancreatin enzyme digest solution 2 h before kidney dissection.
2. Matrigel-treat cell culture plates 1 h before kidney dissection. Thaw Matrigel on ice and dilute 1:25 in cold DMEM/F12 and coat plates. Volumes of Matrigel for coating are 50 μL/well for 96-well, 300 μL/well for 24-well, and 1 mL/well for 6-well. Make sure Matrigel is distributed over the entire surface of the well and allow to sit undisturbed in a laminar flow cabinet for cell culture for at least 1 h.
3. Prepare round bottomed 5 mL tubes to receive dissected kidneys by filling them with 2 mL HBSS.
4. Make appropriate volume of NPEM in a 50 mL conical tube.

### 3.2 Isolation of Nephrogenic Zone Cells (NZCs) from Developing Kidneys

1. Observe sterile technique for all aspects of the dissection to ensure that the primary cells will not become contaminated. Using a stereomicroscope and watchmaker's forceps, dissect the kidneys from E13 to P1 mice in a 10 cm dish containing a sufficient volume of PBS to cover the embryos. Completely remove the ureter and kidney capsule to expose the nephrogenic zone, and transfer kidneys to the HBSS-containing tube using a transfer pipette with an orifice cut to at least twice the diameter of the kidneys to be transferred. Discard ruptured or

broken kidneys as they will reduce the purity of the final NPC preparation. Up to 24 kidneys can be placed in one 5 mL round bottom tube. A demonstration video of the dissection procedure is included in our published method for isolating cells from the nephrogenic zone [7].

2. Incubate the tube with kidneys in HBSS on a nutator for 2 min to dislodge debris that may still be attached to the kidneys (e.g., red blood cells and remaining capsule fragments).
3. Remove the HBSS and add another 2 mL of HBSS to wash. Be careful not to damage the kidneys with the pipette tip.
4. Carefully remove as much of the HBSS as possible, and add 2 mL of RT collagenase A/pancreatin enzyme digest solution. Place the tube on a nutator at 37 °C for 10–16 min (the exact time must be empirically determined for the batch of enzyme solution).
5. After digestion, remove the tube containing kidneys, and immediately add 125 μL of FBS to stop the enzyme reaction. Invert to mix. Perform steps from here on in a laminar flow cabinet for cell culture.
6. Allow kidneys to sink to the bottom of the tube, and remove any floating particles with a 1 mL micropipette while removing as little of the cell suspension as possible. Floating particles at this stage in the purification are usually capsule fragments that were not removed prior to the digestion.
7. Remove remaining cell suspension with a 1 mL micropipette for downstream cell purification. There is no need to take all of the solution, and leaving 200 μL behind is recommended to avoid unwanted particles at the bottom of the tube.
8. Transfer the cell suspension evenly to two 1.5 mL microfuge tubes, and spin in a microfuge at 2000 rpm (300 *g*) for 5 min.
9. Discard the supernatant; gently resuspend each cell pellet in 500 μL PBS-BSA; then combine cell suspensions into one 1.5 mL microfuge tube.
10. Place a 30 μm pre-separation filter on a 15 mL conical tube, and wash with 4 mL of PBS-BSA. Discard the flow through.
11. Add cell suspensions to the washed 30 μm pre-separation filter, and wash with 500 μL of PBS-BSA. Transfer the 1.5 mL of filtered cell suspension to a 1.5 mL microfuge tube.
12. Transfer 10 μL of the cell suspension into a microfuge tube containing 90 μL PBS-BSA, and determine cell count using a hemocytometer.
13. For NPC purification proceed to Subheading 3.4. For a mixed population of NZCs, spin the cells at 300 *g* for 5 min to pellet and resuspend in medium for plating.

### 3.3 Nephron Progenitor Cell (NPC) Purification from Nephrogenic Zone Cells

1. For up to ten million NZCs, gently resuspend cell pellet to a total volume of 76 μL (taking into consideration the volume of pellet) with autoMACS running buffer in a 1.5 mL microfuge tube, and add the following antibodies for a total volume of 110 μL: 9 μL of anti-CD105-PE, 9 μL of anti-CD140-PE, 8 μL of anti-Ter119-PE, and 8 μL of anti-CD326-PE.
2. Triturate cell/antibody suspension gently with a 200 μL pipettor, and incubate at 4 °C for 13 min without agitation.
3. Add 1 mL autoMACS running buffer and triturate gently three times. Centrifuge sample at 300 *g* for 5 min.
4. Discard supernatant and gently resuspend pellet in 1 mL autoMACS running buffer. Centrifuge sample at 300 *g* for 5 min.
5. Discard supernatant and gently resuspend the pellet in a total volume of 80 μL autoMACS running buffer (taking into consideration the volume of pellet).
6. Add 20 μL of anti-PE microbeads to cell suspension. Gently mix by triturating several times, and incubate at 4 °C for 18 min without agitation.
7. Add 1 mL autoMACS running buffer and triturate gently three times. Centrifuge sample at 300 *g* for 5 min.
8. Discard supernatant and resuspend cell pellet in 1 mL autoMACS running buffer. Centrifuge again at 300 *g* for 5 min.
9. Discard supernatant and resuspend the pellet in 1 mL autoMACS buffer.
10. Attach MACS Separator to the MultiStand and place LS Column in the separator. Place a collection tube under the LS Column. Make sure to insert the LS Column with the column wings facing to the front of the MACS Separator.
11. Prepare LS Column by rinsing with 3 mL autoMACS running buffer on top of the column, and let the buffer run through. Discard effluent and change collection tube. The LS Column is now ready for magnetic separation.
12. Apply cell suspension onto the prepared LS Column and collect the effluent. Wash the column three times with 3 mL of autoMACS running buffer per wash, and collect the total effluent. Add each 3 ml buffer only after the column has finished dripping. The total effluent of the initial cell suspension and the three washes represents the unlabeled fraction (NPCs).
13. Spin the negative fraction at 300 *g* for 5 min to pellet the purified NPCs.
14. Remove the supernatant and resuspend the NPCs in 1 mL of NPEM. Use 10 μL of the suspension for a cell count. The

average yield of NPCs recovered from five million NZCs is approximately 2–2.5 million NPCs.

### 3.4 Cell Plating and Culture

1. Suggested cell seeding density is between 5000 and 25,000 cells per $cm^2$. Increase the volume of NPC suspension with NPEM such that the desired cell density per ml is achieved for plating.
2. Immediately prior to plating, remove Matrigel from the culture plate wells using a vacuum aspirator connected to a sterile pipette tip, and immediately add the desired volume of cell suspension in NPEM to each well. Agitate the plate to spread cells evenly.
3. Optional: Several hours after plating when cells have adhered to the culture plate, excess NPEM can be removed and stored at 4 °C for several days to be used for the next passage. Prior to NPEM storage, pellet any dead cells by centrifugation at 300 *g*, and transfer the supernatant to a new tube while leaving pelleted cells or cellular debris behind. For example, to avoid uneven growth due to meniscus formation, cell can be plated in a 2 mL volume in a one well of a 24-well culture plate, and 1.5 mL of NPEM can be removed after 2 h and stored for use in subsequent passages (*see* **Note 5**).
4. Change 100% of the medium every 48 h.
5. When the cells have reached 70–80% confluence at the areas of the well with the highest cell density, they must be passaged for further expansion. The highest cell densities usually occur at the edge of the well but sometimes can occur in the middle of the well. If cells are allowed to become too dense, they will lose the potential for multiple passages. Figure 2 shows the optimal density that cells should be allowed to grow to before passage.
6. To passage, warm autoMACS running buffer to RT and TrypLE dissociation solution to 37 °C.
7. Aspirate NPEM and add pre-warmed TrypLE solution in the following amounts based upon well size: 96-well, 50 μL; 24-well, 250 μL; and 6-well, 500 μL.
8. Place culture plate in a 37 °C incubator for 2 min, and then check that the cells have started to dislodge from the plate surface under a light microscope. If not, extend the treatment in 1 min increments until cell detachment becomes noticeable.
9. Add the following volumes of autoMACS running buffer to each well, and wash cells from plate by triturating gently: 96-well, 200 μL; 24-well, 1000 μL; and 6-well, 2000 μL.
10. Transfer the cells to an appropriate tube, and spin at 300 *g* for 5 min to pellet the cells.

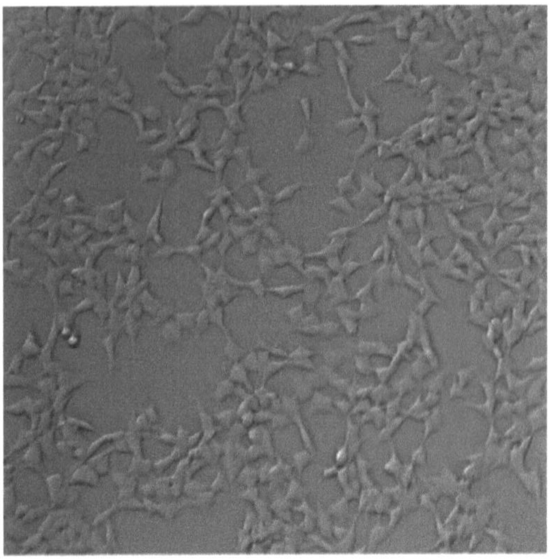

**Fig. 2** Optimal density of NPCs for passaging

11. Remove the supernatant and resuspend the NPCs in 1 mL of NPEM. Perform a cell count on a 1:10 dilution of cells.
12. Increase the volume of NPEM, and plate cells at the desired density on Matrigel-coated culture plates.

### *3.5 Cryopreserving NPCs*

1. NPCs can be frozen in CryoStem Freezing Medium immediately after isolation (uncultured) or following propagation and passaging in NPEM. 100,000 to one million cells per milliliter cryopreservation medium per cryovial is recommended.
2. Pellet uncultured cells or TrypLE-dissociated cells that have been diluted in autoMACS running buffer by centrifugation at 300 *g*, and remove supernatant. If cryopreserving cultured cells, wash one additional time in autoMACS running buffer prior to adding freeze medium to ensure that all TrypLE has been removed. Cells have been successfully cryopreserved from 80% confluent cultures in 24-well and 6-well plates up to passage 5.
3. Resuspend cell pellet in an appropriate volume of chilled cryopreservation medium depending on the number of cells per vial needed, and transfer cell suspension to screw cap cryovials.
4. Place cryovial containing cells in a Styrofoam tube holder with cover at −70 °C overnight to slowly freeze the cells. The next day transfer the cryovial to liquid nitrogen for long-term storage.
5. Thaw cells rapidly in a 37 °C water bath by swirling until just a small sliver of ice is left in the cryovial. This takes approximately 90–120 s.

6. Triturate once with 1 mL micropipette and transfer to a 15 mL conical tube. Slowly drip 2 mL RT autoMACS running buffer into the thawed cell suspension while simultaneously swirling to mix. After 2 mL has been added by dripping, add additional autoMACS running buffer to 10 mL. Invert tube to ensure complete mixing of cryopreservation medium; otherwise, cells may fail to pellet.
7. Spin cell suspension at 300 *g* for 5 min to pellet cells.
8. Remove the majority of autoMACS running buffer (leave 50 μL behind to avoid disrupting the pellet), and wash pellet with 1 mL of NPEM by triturating gently several times.
9. Transfer cells to a 1.5 mL microfuge tube, and spin in a microfuge for 5 min at 300 *g*.
10. Discard NPEM and resuspend cell pellet in the appropriate volume of fresh NPEM for plating. It is important that cells do not clump as they will then cluster, causing focal overgrowth.
11. Plate cells on Matrigel-coated culture plates. We typically observe greater than 90% recovery of viable cells after thawing.

## 4 Notes

1. Our laboratory has used the NPC culture method regularly since it was first published. The method is robust, but we have documented a few avoidable problems over this time: (1) It is important not to use the labeled antibodies for cell depletion beyond their expiration dates. (2) AutoMACS buffer is delivered in opaque 1 L bottles, and it is not possible to see contamination or particulate matter; for this reason we aliquot the autoMACS buffer into clear 50 mL tubes and store at 4 °C. (3) We recommend using rho kinase inhibitor suspended in DMSO (Fig. 3). Our attempts to use rho kinase inhibitor suspended in water have been unsuccessful, and we see extensive clumping of cells, which is characteristic of inefficient rho kinase inhibition. (4) We recommend using baculovirus-produced FGF9 (Fig. 4). Attempts to use *E. coli*-produced FGF9 at the same concentration were unsuccessful; interestingly, some SIX2 expression is maintained in NPCs indicating that it functions to some degree in this procedure, but it yields very poor cell growth.
2. APEL medium was formulated by Dr. Andrew Elefanty's group for the culture of human embryoid bodies [8]. Because it was used as the basal medium for directed differentiation of human pluripotent stem cells to kidney progenitor cells [9], we selected it as the basal medium for NPC culture to enable the

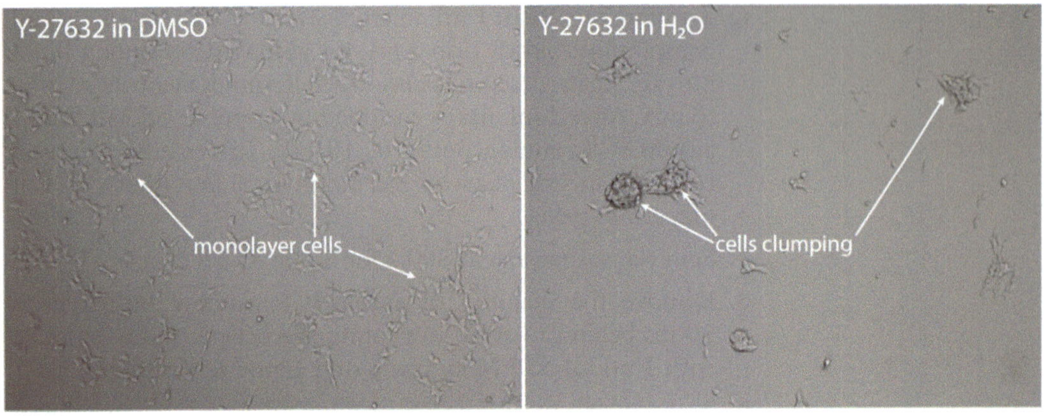


**Fig. 3** Comparison of NPCs cultured in NPEM with Rho kinase inhibitor in DMSO versus water. The small molecule inhibitor can be purchased in solution in either of these vehicles

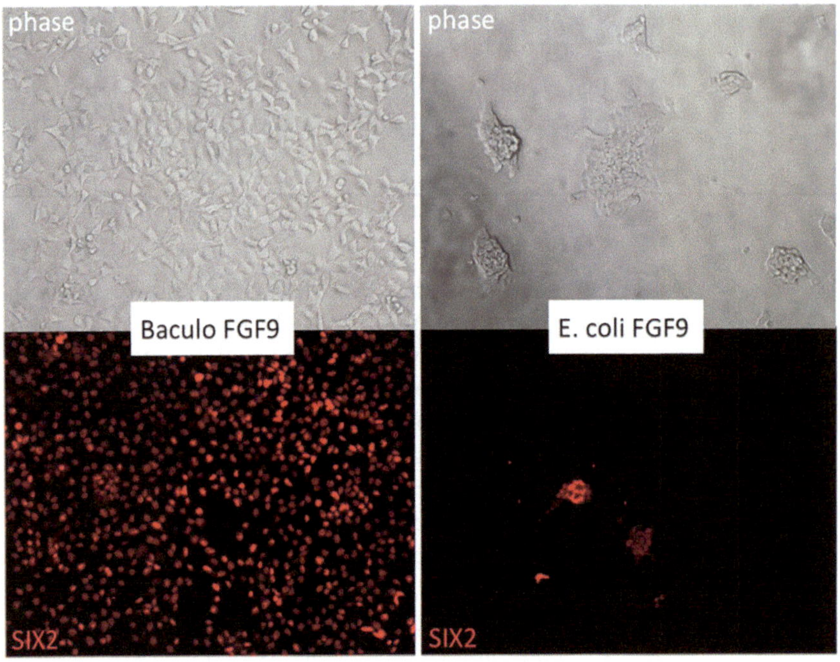


**Fig. 4** Comparison of NPCs grown in NPEM with baculovirus-produced FGF9 and *E. coli*-produced FGF9

use with both mouse and human cells [6]. However, the vendor for APEL, Stemcell Technologies, has now replaced it with APEL 2 and maintains that APEL 2 supplemented with PFHM serves as an equivalent substitution. Several investigators using the protocol for mouse NPC propagation described in [6] have reported problems reproducing the conditions with APEL 2 + PFHM (personal communications). In our hands, NPCs can be propagated using APEL 2 + PFHM as a substitute for APEL if we adhere strictly to the protocol reported here

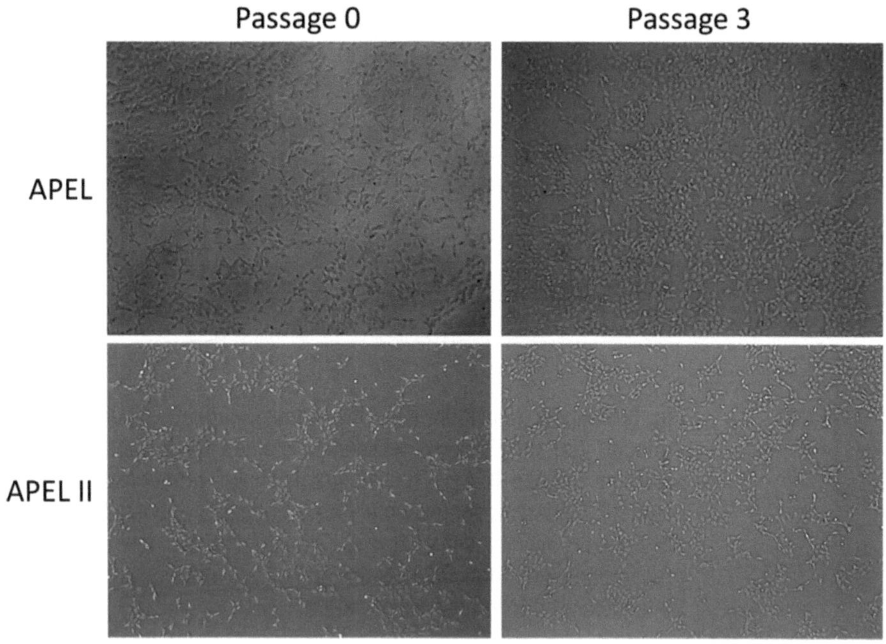


**Fig. 5** Comparison of NPC culture in APEL- and APEL2 + PFHM-based NPEM. Cells are shown at two passages: first outgrowth on plate after isolation (passage 0) and passage 3

(Fig. 5). However, the morphology of cells is slightly different from our initial cultures with APEL, and cells cultured in NPEM based on APEL 2 + PFHM show more cytoplasmic projections, leading us to believe that APEL 2 + PFHM is not fully equivalent to APEL. We have observed that the tolerance for variability in quality of the other components of NPEM is much lower with APEL 2 + PFHM than with APEL, and it is therefore very important that all other factors have full biological activity.

3. We have found that the purity of cells isolated using the Miltenyi manual cell purification described above is over 99% (Fig. 6). For use of an automated cell purifier for this procedure, please refer to [5]. Yields of approximately 95% can be achieved with the Miltenyi autoMACS system.
4. All growth factors and small molecule inhibitors should be stored frozen at −20 °C. Once thawed, they are stable for up to 2 weeks at 4 °C. Dilutions of IGF1, IGF2, and LDN-193189 in APEL medium should be made fresh during each preparation of NPEM. NPEM should be made fresh and used within 1 week.
5. To prevent clustering of cells near the edge of the well, which will result in earlier cellular overgrowth and the need for more frequent passaging, it is highly recommended to initially plate the cells in a high volume of NPEM. This will reduce the effect

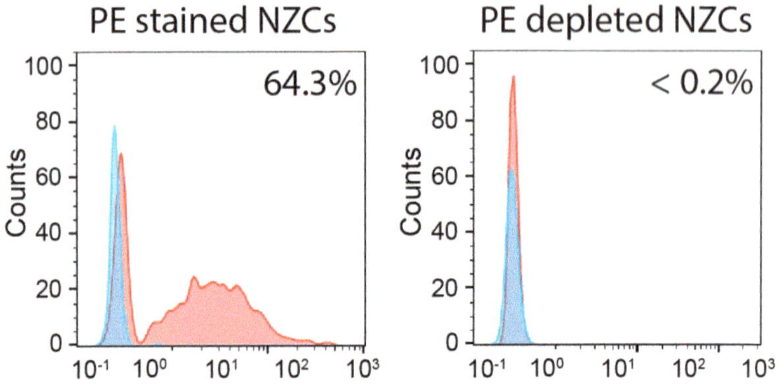


**Fig. 6** Magnetic depletion using the manual magnetic column described in the protocol. The blue peak represents unstained NPCs, and the orange peaks represent stained NPCs. The left graph shows immunostained cells prior to magnetic depletion: approximately 65% of cells are labeled with antibody, and 35% of cells overlap with unlabeled. The graph to the right shows the same cell preparation after magnetic depletion: only 0.2% of the cells are now labeled, and the unlabeled fraction represents NPCs

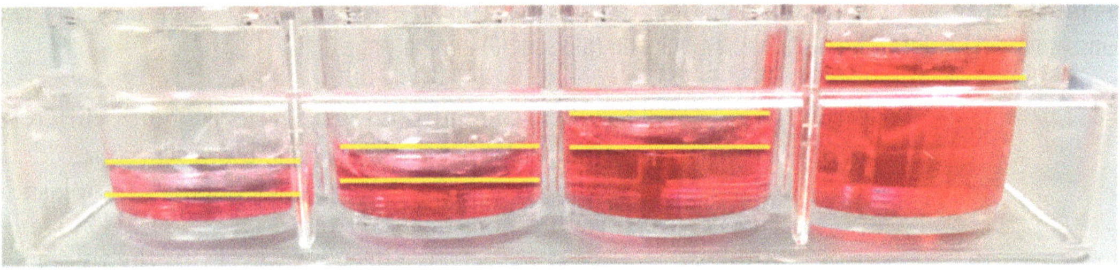

**Fig. 7** Meniscus formation in a 24-well plate containing different volumes of medium

that the meniscus has on cell distribution on the culture dish surface (Fig. 7). The following volumes can be used as a guide: 96-well plate, 200 μL; 24-well plate, 2 mL; 6-well plate, 4 mL; and 10 cm dish, 10 mL.

## Acknowledgments

Research reported in this publication is supported by the National Institutes of Health/National Institute of Diabetes and Digestive and Kidney Diseases under awards R01DK078161 (L.O.) and R24DK106743 (L.O.). Core facilities support was provided by the Maine Medical Center Research Institute core facilities for Molecular Phenotyping and Progenitor Cell Analysis [supported by the National Institute of General Medical Sciences (NIGMS) 5P30 GM106391]. The content is solely the responsibility of the authors and does not necessarily represent the official views of the National Institutes of Health.

## References

1. Kobayashi A, Valerius MT, Mugford JW, Carroll TJ, Self M, Oliver G, McMahon AP (2008) Six2 defines and regulates a multipotent self-renewing nephron progenitor population throughout mammalian kidney development. Cell Stem Cell 3(2):169–181. https://doi.org/10.1016/j.stem.2008.05.020
2. Boyle S, Misfeldt A, Chandler KJ, Deal KK, Southard-Smith EM, Mortlock DP, Baldwin HS, de Caestecker M (2008) Fate mapping using Cited1-CreERT2 mice demonstrates that the cap mesenchyme contains self-renewing progenitor cells and gives rise exclusively to nephronic epithelia. Dev Biol 313(1):234–245. https://doi.org/10.1016/j.ydbio.2007.10.014
3. Blank U, Brown A, Adams DC, Karolak MJ, Oxburgh L (2009) BMP7 promotes proliferation of nephron progenitor cells via a JNK-dependent mechanism. Development 136(21):3557–3566. https://doi.org/10.1242/dev.036335
4. Brown AC, Adams D, de Caestecker M, Yang X, Friesel R, Oxburgh L (2011) FGF/EGF signaling regulates the renewal of early nephron progenitors during embryonic development. Development 138(23):5099–5112. https://doi.org/10.1242/dev.065995
5. Brown AC, Muthukrishnan SD, Guay JA, Adams DC, Schafer DA, Fetting JL, Oxburgh L (2013) Role for compartmentalization in nephron progenitor differentiation. Proc Natl Acad Sci U S A 110(12):4640–4645. https://doi.org/10.1073/pnas.1213971110
6. Brown AC, Muthukrishnan SD, Oxburgh L (2015) A synthetic niche for nephron progenitor cells. Dev Cell 34(2):229–241. https://doi.org/10.1016/j.devcel.2015.06.021
7. Brown AC, Blank U, Adams DC, Karolak MJ, Fetting JL, Hill BL, Oxburgh L (2011) Isolation and culture of cells from the nephrogenic zone of the embryonic mouse kidney. J Vis Exp 50. https://doi.org/10.3791/2555
8. Ng ES, Davis R, Stanley EG, Elefanty AG (2008) A protocol describing the use of a recombinant protein-based, animal product-free medium (APEL) for human embryonic stem cell differentiation as spin embryoid bodies. Nat Protoc 3(5):768–776. https://doi.org/10.1038/nprot.2008.42
9. Takasato M, Er PX, Becroft M, Vanslambrouck JM, Stanley EG, Elefanty AG, Little MH (2014) Directing human embryonic stem cell differentiation towards a renal lineage generates a self-organizing kidney. Nat Cell Biol 16(1):118–126. https://doi.org/10.1038/ncb2894

# Chapter 7

# 3D Cell Culture Models of Epithelial Tissues


## Kai Zhang and Aki Manninen



## Abstract

Cells in tissues in vivo face a very different microenvironment than typical cultured cells plated on a plastic dish. Already several decades ago, cell biologists observed that cell lines show dramatically different morphology and growth characteristics when embedded into three-dimensional (3D) substrates or standard tissue culture plates (Montesano R, Schaller G, Orci L, Cell. 66:697–711, 1991; Barcellos-Hoff MH, Aggeler J, Ram TG, Bissell MJ, Development. 105:223–235, 1989; Simian M, Bissell MJ, J Cell Biol. 216:31–40, 2017). Despite its imminent benefit for cell biological studies, suspicion and prejudice toward more complicated sample preparation requirements limited the popularity of 3D culture techniques until recently, when it was shown that soft 3D gels made of basement membrane extracts (BME) allow prolonged culture of many types of primary epithelial cells (Clevers H, Cell. 165:1586–1597, 2016; Sato T, Vries RG, Snippert HJ, van de Wetering M, Barker N, Stange DE, et al., Nature. 459:262–265, 2009). These observations have brought 3D organoid culture systems into the mainstream. Here we describe two protocols for culturing epithelial cells in 3D substrates, the "blob culture" setup where cells are fully embedded into BME gel and the "overlay setup" where cells are seeded on top of BME gel and then overlaid with a thin layer of BME (Debnath J, Brugge JS, Nat Rev Cancer. 5:675–688, 2005; Bryant DM, Datta A, Rodriguez-Fraticelli AE, Peranen J, Martin-Belmonte F, Mostov KE, Nat Cell Biol. 12:1035–1045, 2010).



**Key words** MDCK, 3D culture, Organoid


## 1 Introduction

Most of the early 3D culture work was based on Madin-Darby canine kidney (MDCK) and various mammary epithelial cell lines such as MCF10A [1, 2]. Here we describe a protocol developed for 3D Matrigel® culture of MDCK cells, but the protocol is also applicable to most other epithelial cell lines [3–6]. Obviously, the mixture of growth and other factors that need to be added to the culture medium must be optimized separately for different cell lines and are of particular importance when culturing primary cell organoids [7, 8]. The selection of the appropriate method from the two variations of the protocol depends on the analysis to be made. The blob setup allows growth of higher number of cells, and they can be


Seppo Vainio (ed.), *Kidney Organogenesis: Methods and Protocols*, Methods in Molecular Biology, vol. 1926,
https://doi.org/10.1007/978-1-4939-9021-4_7,

mounted for high-resolution imaging using confocal microscopy but are less accessible for high-resolution live cell imaging [9]. The overlay setup is more suitable for live cell imaging studies and readily accessible for certain manipulations, such as viral transduction of organoid cultures, during culture [10].

## 2 Materials

Prepare all solutions intended for cell culture using sterile ultrapure water, and work aseptically in a laminar flow cabinet. All materials should be cell culture tested and of high purity.

### 2.1 MDCK Cell Culture

1. 3D complete culture medium: MEM-GlutaMAX supplied with 10% fetal bovine serum (FBS) and 1% penicillin/streptomycin.
2. 2D complete culture medium: As above but with 5% FBS.
3. Trypsin-EDTA.
4. Matrigel® matrix (BD Biosciences): Thaw on ice, mix well, and aliquot into microtubes and store at −80 °C.
5. 35 mm Ø glass-bottom μ-Dish (IBIDI): Cool it on ice before use.

### 2.2 Immunofluorescence (IF)

1. Fixation buffer: 3–4% (para) formaldehyde (PFA) in PBS (0.1% glutaraldehyde may need to be added, *see* **Note 1**).
2. Fixative quenching buffer: 100 mM glycine in PBS.
3. Permeabilization buffer: 0.2% TX-100 in PBS.
4. IF Buffer: 130 mM NaCl, 7 mM $Na_2HPO_4$, 3.5 mM $NaH_2PO_4$, 7.7 mM $NaN_3$, 0.1% BSA, 0.2% Triton X-100, 0.05% Tween-20.
5. Nuclear dye: 1 mg/mL of 4′,6-diamidino-2-phenylindole (DAPI, Sigma) in PBS.
6. Mounting medium: Immu-Mount (Thermo Scientific).
7. A confocal microscope with appropriate image analysis software, such as Zeiss LSM 780 confocal microscope, 40× Plan-Apochromat objective (N.A. = 1.4), and the ZEN 2011 software.

### 2.3 Live Cell Imaging

1. A fluorescent protein-based marker to label cells such as mRed-fused to histone 2B to stain cell nuclei (pHIV-H2B-mRed lentivirus vector [11] was used in this example). Cell may be sorted prior to imaging (*see* **Note 2**).
2. A sensitive spinning disk confocal microscopy setup with environmental control and suitable image analysis software such as Zeiss Cell Observer with a spinning disk unit (Yokogawa CSU-X1) and Hamamatsu EMCCD camera equipped with

humidified, environmental control incubator (37 °C, 5% $CO_2$) and Zen 2012 (Blue edition) software.

## 3 Methods

### 3.1 3D Overlay Culture of MDCK Cells

1. Split nearly confluent MDCK cells 1 day before 1:5. This helps to avoid cell clumping that is detrimental for proper growth of MDCK cysts in 3D gels.
2. Thaw Matrigel® on ice (allow 1–2 h for small aliquots (<1 mL) and overnight for larger aliquots (1–10 mL)). *See* **Note 3**.
3. Add 150–200 μL Matrigel® onto a 35 mm high μ-dish. Spread Matrigel® evenly with a 200 μL tip to reduce possible meniscus forming at the edges of the glass-bottom dish (*see* **Note 4**). Take care not to generate air bubbles. Place the dish at +4 °C for 5 min.
4. Transfer the dish into +37 °C, $CO_2$ incubator, and allow 20 min for the Matrigel® to solidify.
5. Wash cells (~40–60% confluent) with PBS, add 2 mL of Trypsin-EDTA, and incubate for 10 min or until cell detaches at +37 °C. Resuspend trypsinized cells into 8 mL of 2D complete culture medium.
6. Count cells and aliquot 60,000 cells into a fresh 15 mL conical tube; fill to 10 mL with 2D complete culture medium.
7. Tube A (cell suspension): Pellet cells at 150 × *g* for 3 min. Resuspend cell pellet into 5 mL of 3D culture medium (label this tube "A"). This will generate 1200 cells/100 μL; 750 μL or 9000 cells will be seeded per dish. *See* **Note 5**.
8. Tube B (diluted Matrigel®): Prepare 3D 750 μL per sample of 3D culture medium containing 4% Matrigel® (label this tube "B"). Use ice-cold medium to prevent premature gelation of Matrigel®.
9. Mix 750 μL of Tube A and 750 μL of Tube B, and plate the mixture onto Matrigel®-coated μ-Dish prepared in **step 3**. Gently transfer the dish into the $CO_2$ incubator.
10. Exchange media every 2 days with fresh culture medium containing 2% Matrigel®, and culture for 6 days. If the culture is continued for prolonged times, you can omit the Matrigel® from the fresh medium after 6th day.

### 3.2 Time-Lapse Imaging of MDCK Cells in 3D Overlay Cultures

1. MDCK cells stably expressing H2B-mRed were set up for 3D culture in 35 mm glass-bottom μ-Dish (Ibidi) as described in Subheading 3.1. And cultured for 72 h in a normal humidified $CO_2$ incubator at +37 °C.

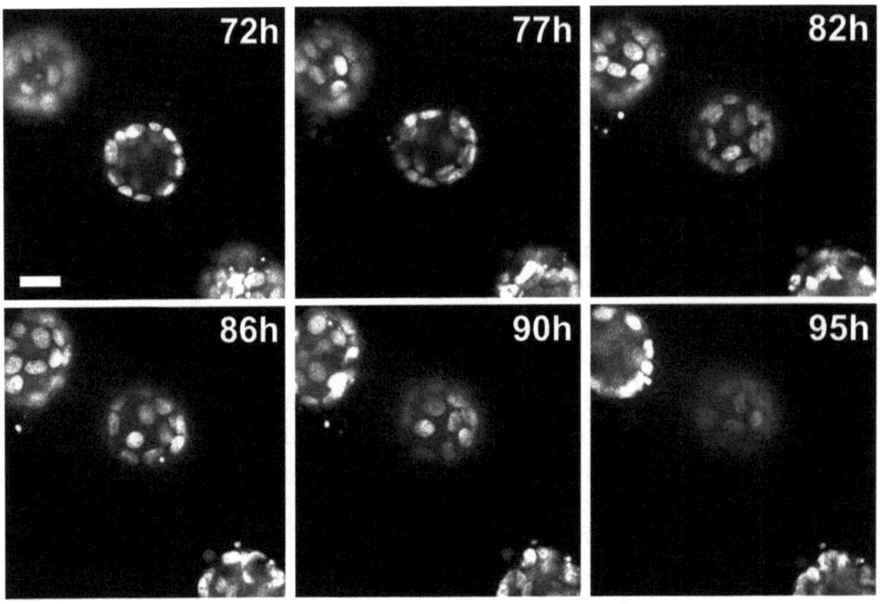


**Fig. 1** Time-lapse imaging of 3D overlay culture of MDCK cells. H2B-mRed expressing MDCK cells were first cultures in 3D overlay setup for 72 h and then placed into an incubation chamber at a spinning disk confocal microscope and imaged at 30 min intervals as described in Subheading 3.2. Selected time points are shown. The scale bar is 2 μm

2. Live imaging was initiated by placing the μ-Dish into Zeiss spinning disk confocal microscope equipped with a spinning disk unit (Yokogawa CSU-X1), Hamamatsu EMCCD camera, EC Plan NeoFluar 40×/0.75 DIC (dry/air) objective and humidified, environmental control incubator (37 °C, 5% $CO_2$).
3. Multipoint imaging was performed by capturing image stacks of complete cysts every 30 min for a duration of 27 h using Zen 2012 software (Blue edition; Fig. 1).

### 3.3 3D Blob Culture of MDCK Cells

1. Split nearly confluent MDCK cells 1 day before 1:5. This helps to avoid cell clumping that is detrimental for proper growth of MDCK cysts in 3D gels.
2. Thaw sufficient amount of Matrigel® on ice (e.g., ~150 μL per sample with three replicates).
3. Chill sterile Eppendorf tubes (one per each sample) on ice (preferably a metal block that can be easily transferred into laminar hood when seeding the cysts).
4. Once thawed, pipette 150 μL of Matrigel® into each of the chilled tubes—Keep cold—If you have problems with Matrigel® gelling too fast, use pre-chilled tips in all steps that involve Matrigel® (*see* **Note 6**).
5. Wash cells (~40–60% confluent) with PBS, add 2 mL of trypsin-EDTA, and incubate for 10 min or until cell detaches

at +37 °C. Resuspend trypsinized cells into 8 mL of 2D complete culture medium.

6. Count cells, and aliquot one million cells into a fresh 15 mL conical tube; fill to 10 mL with 2D complete culture medium.
7. Pellet cells at 150 × *g* for 3 min. Resuspend cell pellet into 400 μL of 3D culture medium (1 × $10^6$ cells/mL).
8. Resuspend cells, and transfer 10 μL of cell suspension (25,000 cells) into Matrigel® tube; mix swiftly by pipetting but avoid air bubbles.
9. Using a 200 μL pipette, gently mix twice and dispense three ~50 μL blobs each into the center of a well in 24-well plate—Avoid contact with the sides of the well, and be careful not to make air bubbles. If air bubbles form, you can try to remove them with a pipette before the gel solidifies.
10. When you have pipetted all of the blobs for all of the samples, transfer the plate into +37 °C $CO_2$ incubator for 30 min to allow solidification and then add 1.5 mL of 3D culture medium per well.
11. Grow cells for 4–10 days and change the growth medium every 2–3 days. Fully polarized MDCK cysts form in 4–7 days.

### 3.4 Immunofluorescence and Preparation of Blob Cultures for Confocal Imaging

1. For immunofluorescence analysis, the blobs are fixed in the wells by pipetting 1 mL of PFA (+0.1% glutaraldehyde) to the side of the wells and incubating for 20 min at room temperature.
2. If necessary, cells are permeabilized by incubating with permeabilization buffer for 30 min followed by incubation for 30 min in quenching buffer, both at room temperature.
3. Nonspecific binding sites are blocked by a 1-h incubation within blocking buffer at room temperature.
4. Primary antibodies: Incubate with primary antibodies in IF buffer (use two- to fourfold concentrations compared with normal immunofluorescence in 2D) overnight at +4 °C, carefully rinse once with 1.5 mL of IF buffer, and perform three 20 min washes with IF buffer at room temperature with gentle rocking.
5. Secondary antibodies: Incubate with 1:400 dilution of fluorophore-conjugated secondary antibody in IF buffer for 4 h at room temperature, or overnight at +4 °C. If nuclear or actin cytoskeletal staining is needed, add DAPI and fluorophore-conjugated phalloidin, respectively, at this step together with secondary antibodies. Rinse once with PBS and wash three times, 20 min each as above. Samples are mounted by pipetting them onto a slide with tape spacers with an

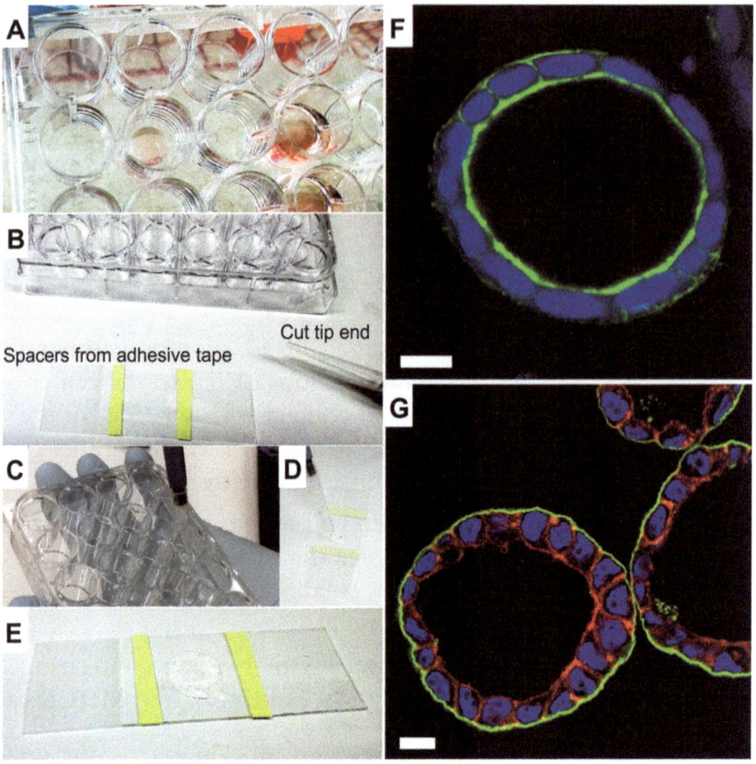


**Fig. 2** Mounting and confocal microscopy of blob cultures. (**a**) Blobs are fixed and stained in 24-well plate wells as described in Subheading 3.4. (**b–d**) Pieces of adhesive tape are used as spacers on slides and tip end is cut with scissors to allow easy pipetting of Matrigel® onto the slide. (**e**) Matrigel is overlaid mounting medium and gently flattened by placing a cover slip onto spacers. (**f**) MDCK cysts were grown for 6 days as described in Subheading 3.3. and stained for filamentous actin using Alexa Fluor 488®-phalloidin (green; A123 79, Thermo Fischer Scientific). Nuclei were stained with DAPI (blue). (**g**) Many types of epithelial cells form hollow cysts in 3D Matrigel® cultures. RWPE-1 human prostate epithelial cells were grown for 6 days as described above (except for using Keratinocyte-SFM; Cat.# 17005042, Thermo Fischer Scientific). Cells were stained for α6-integrin that marks basal surface of epithelial cells (green; GoH3/Abcam #ab 105669), filamentous actin (red; TRITC-phalloidin, P1951, Sigma) and nuclei (blue; DAPI). Scale bar in (**f**) and (**g**) is 10 μm

end-cut 200 μL tip, overlaid with 100 μL of Immu-Mount, and a coverslip (Fig. 2, *see* **Note 7**).

6. Confocal laser scanning microscopy (LSM) images are best acquired by using confocal microscopy such as Zeiss LSM 780 confocal microscope with a 40× Plan-Apochromat objective (N.A. = 1.4). Roughly 30 confocal stacks were acquired with 400 nm increments from the basal region of cysts containing matrix adhesions using the ZEN 2011 software.

## 4 Notes

1. 0.1% glutaraldehyde may be added into PFA-fixation solution if your Matrigel® blobs tend to dissolve during PFA-fixation step.
2. *Optional Step*: Fluorescence-activated cell sorting (FACS) setup such as Becton Dickinson FACSAria™ III Cell Sorter may be used in order to enrich for positively transduced cells.
3. Always keep Matrigel® on ice; it will solidify at room temperature. Thaw Matrigel on ice and do not warm the tubes in hands.
4. Use a 200-μL tip to spread the Matrigel® evenly to the bottom of μ-Dish and to remove any possible air bubbles that would interfere with imaging. Do not spend too much time at this step as small amounts of Matrigel® will warm up quickly.
5. Precool all the tubes and tips that you use for handling liquid Matrigel®. Matrigel will start to polymerize when temperature reaches ~ +10 °C.
6. Important! Do pipetting gently but swiftly; try to keep all the Matrigel® mixtures ice-cold (chilled metal block in the laminar) to prevent premature gel formation.
7. To minimize the breaking of the Matrigel®, use a 1 mL end-cut tip for the Matrigel® transfer onto cover slip.

## Acknowledgments

Riitta Jokela is acknowledged for overall expert technical assistance, Jaana Träskelin for expert technical assistance at Biocenter Oulu Virus Core Laboratory, and Veli-Pekka Ronkainen for expert assistance in microscopy at Biocenter Oulu Tissue Imaging Center. This work was funded by the Academy of Finland (251314, 135560, 263770, and 140974/AM).

### References

1. Montesano R, Schaller G, Orci L (1991) Induction of epithelial tubular morphogenesis in vitro by fibroblast-derived soluble factors. Cell 66:697–711
2. Barcellos-Hoff MH, Aggeler J, Ram TG, Bissell MJ (1989) Functional differentiation and alveolar morphogenesis of primary mammary cultures on reconstituted basement membrane. Development 105:223–235
3. Debnath J, Brugge JS (2005) Modelling glandular epithelial cancers in three-dimensional cultures. Nat Rev Cancer 5:675–688
4. Bryant DM, Datta A, Rodriguez-Fraticelli AE, Peranen J, Martin-Belmonte F, Mostov KE (2010) A molecular network for de novo generation of the apical surface and lumen. Nat Cell Biol 12:1035–1045
5. Simian M, Bissell MJ (2017) Organoids: a historical perspective of thinking in three dimensions. J Cell Biol 216:31–40
6. Manninen A (2015) Epithelial polarity—generating and integrating signals from the ECM with integrins. Exp Cell Res 334:337–349

7. Clevers H (2016) Modeling development and disease with organoids. Cell 165:1586–1597
8. Sato T, Vries RG, Snippert HJ, van de Wetering M, Barker N, Stange DE et al (2009) Single Lgr5 stem cells build crypt-villus structures in vitro without a mesenchymal niche. Nature 459:262–265
9. Myllymäki SM, Teräväinen TP, Manninen A (2011) Two distinct integrin-mediated mechanisms contribute to apical lumen formation in epithelial cells. PLoS One 6:e19453
10. Zhang K, Myllymaki SM, Gao P, Devarajan R, Kytola V, Nykter M et al (2017) Oncogenic K-Ras upregulates ITGA6 expression via FOSL1 to induce anoikis resistance and synergizes with αV-class integrins to promote EMT. Oncogene 36:5681–5694
11. Welm BE, Dijkgraaf GJ, Bledau AS, Welm AL, Werb Z (2008) Lentiviral transduction of mammary stem cells for analysis of gene function during development and cancer. Cell Stem Cell 2:90–10

# Part III

## Programming the Kidney Cell Fates in Human Pluripotent Cells

# Chapter 8

# Generation of Three-Dimensional Nephrons from Mouse and Human Pluripotent Stem Cells

Yasuhiro Yoshimura, Atsuhiro Taguchi, and Ryuichi Nishinakamura

## Abstract

Nephrons, the functional units of the kidney, are derived from nephron progenitor cells (NPCs). Here, we describe methods to reconstruct nephron tissue via induction of NPCs from mouse and human pluripotent stem cells, which mimic multistep developmental signals in vivo. Induced NPCs differentiate into three-dimensional nephron structures, including glomerular podocytes and nephric tubules, which are useful for studying early stages of kidney specification and morphogenetic processes in the context of normal development or disease.

**Key words** Kidney, Nephron, Regeneration, Pluripotent stem cells, Mouse, Human, Nephron progenitor cells, Growth factors, Flow cytometry

## 1 Introduction

As functional units of the kidney, nephrons produce urine by blood filtration, reabsorption, and excretion mechanisms. Impairment of nephrons leads to many types of kidney diseases, such as glomerulonephritis and renal tubular injury. Thus, research involving de novo nephron induction from pluripotent stem cells has been desired to study kidney diseases and realize new therapies.

The mammalian kidney develops from two different progenitor populations: metanephric mesenchyme (MM) and ureteric bud (UB) [1]. Nephron progenitor cells (NPCs), the progenitor cells that give rise to nephrons, exist in the MM, whereby they differentiate into glomerular epithelia and renal tubular epithelia in response to Wnt signaling from the UB [2]. However, from where NPCs originate and how they develop from this origin had not been elucidated. This was a major barrier to establishing proper methods for inducing NPCs from pluripotent stem cells.

In 2013, we revealed that mouse MM develops from *T*(Brachyury)-positive cells within the caudal portion of embryonic day 8.5 (E8.5) embryos via the posterior intermediate mesoderm of

Seppo Vainio (ed.), *Kidney Organogenesis: Methods and Protocols*, Methods in Molecular Biology, vol. 1926,
https://doi.org/10.1007/978-1-4939-9021-4_8, © Springer Science+Business Media, LLC, part of Springer Nature 2019

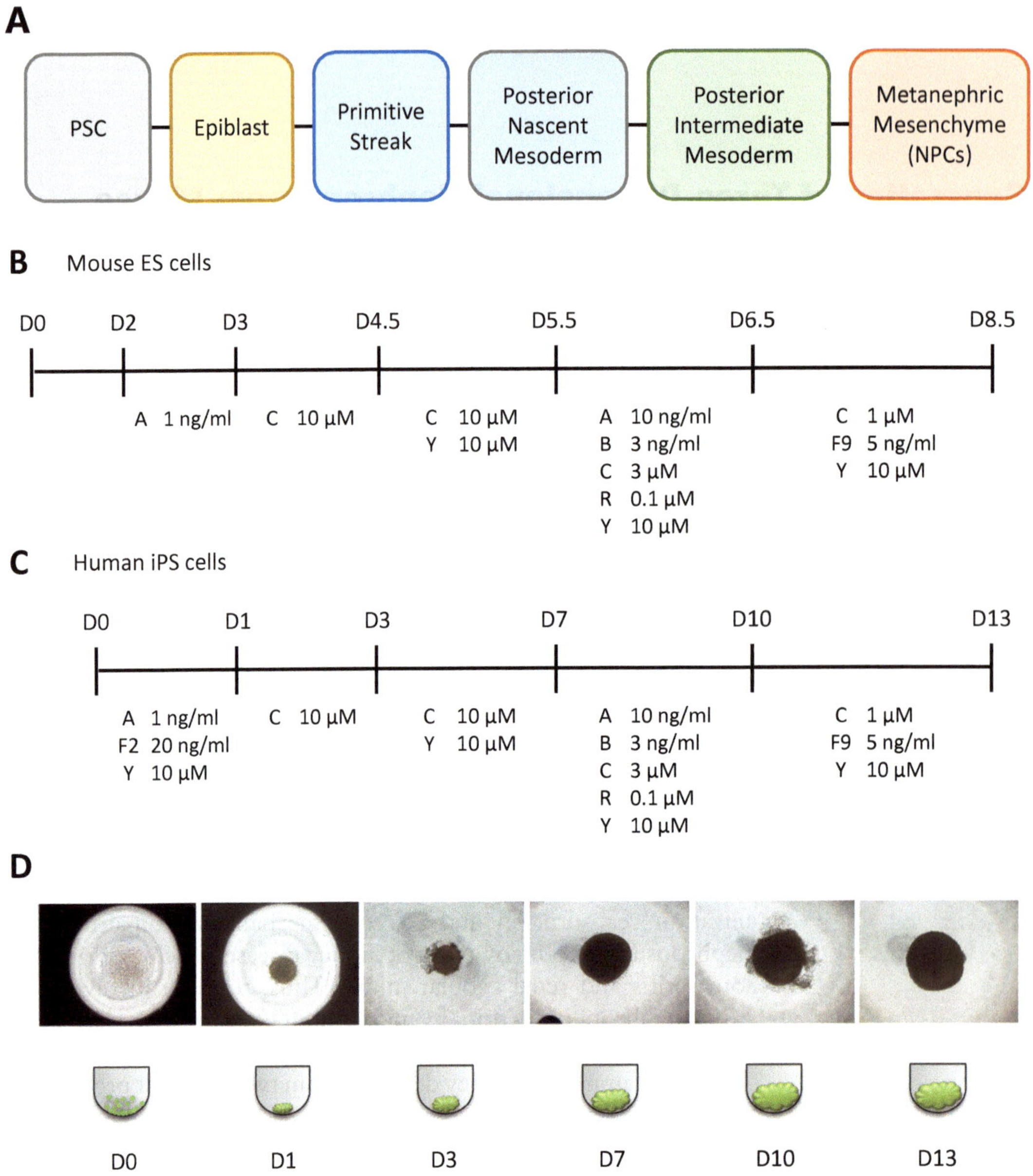


**Fig. 1** Protocols for NPC induction from mouse ES cells and human iPS cells. (**a**) Developmental progression toward NPCs. (**b**, **c**) Outline of protocol for NPC induction from mouse ES cells (**b**) and human iPS cells (**c**). (**d**) Time-course photos and schematic diagrams of NPC induction from human iPS cells. Human iPS cells aggregate to form small spheres at day 1. Spheres gradually become larger with time. *PSC* pluripotent stem cell, *NPCs* nephron progenitor cells, *A* activin A, *B* BMP4, *C* CHIR99021, *F2* Fgf2, *F9* Fgf9, *R* Retinoic acid, *Y* Y27632

E9.5 embryos. These findings about the origin of NPCs in early-stage embryos enabled us to establish a method to induce NPCs from pluripotent stem cells (Fig. 1a–c), which readily reconstituted the three-dimensional structure of nephrons [3]. Recently, we further optimized this protocol to achieve higher induction efficiency of NPCs by introducing a flow cytometric quantification system to monitor the integrin subunit alpha 8-positive/platelet-derived growth factor receptor-negative ($ITGA8^+/PDGFR^-$) NPC population [4].

In this chapter, we provide detailed protocols for induction of NPCs from mouse embryonic stem (ES) cells and human-induced pluripotent stem (iPS) cells, analysis of NPCs by flow cytometry, and generation of three-dimensional nephrons structured by co-culturing NPCs with mouse embryonic spinal cord. These methods are applicable to study kidney development [5, 6], organogenesis [4], and disease both in mouse and human contexts.

## 2 Materials

1. Mouse ES cells (*see* **Note 1**).
2. Human iPS cells (*see* **Note 2**).
3. Pregnant mice: Mouse embryonic age is based on the day of the appearance of a vaginal plug in a pregnant mouse, which is defined as E0 (*see* **Note 3**).
4. Mouse ES cell maintenance medium: Dulbecco's Modified Eagle's Medium (DMEM; Thermo Fisher Scientific, Cat# 11995-065) supplemented with 15% fetal bovine serum (FBS; Sigma-Aldrich, Cat# 172012), 1% (v/v) nonessential amino acids (Thermo Fisher Scientific, Cat# 11140-050), 0.1 mM 2-mercaptoethanol (Nacalai Tesque, Cat# 21438-82), and 1000 U/mL leukemia inhibitory factor (LIF; EMD Millipore, ESG1107).
5. 2i medium: DMEM supplemented with 15% FBS, 1% (v/v) nonessential amino acids, 0.1 mM 2-mercaptoethanol, 1000 U/mL LIF, 3 μM CHIR99021 (Axon Medchem, Cat# Axon1386), and 1 μM PD0325901 (Wako, Cat# 163-24001) (*see* **Note 4**).
6. Serum-free differentiation medium for mouse ES cells: 75% Iscove's Modified Dulbecco's Medium (IMDM; Thermo Fisher Scientific, Cat# 21056-023) and 25% Ham's F12 medium (Thermo Fisher Scientific, Cat# 11765-054) with 0.5× N2 (Thermo Fisher Scientific, Cat# 17502-048), 0.5× B27 without retinoic acid (Thermo Fisher Scientific, Cat# 12587-010), 0.5× penicillin/streptomycin (Thermo Fisher Scientific, Cat# 15070-063), 0.05% bovine serum albumin (BSA) (Sigma-Aldrich, Cat# A4503), 2 mM L-glutamine (Thermo Fisher Scientific, Cat# 25030-081), 0.5 mM ascorbic acid (Sigma-Aldrich, Cat# A4403), and $4.5 \times 10^{-4}$ M 1-thioglycerol (Sigma-Aldrich, Cat# M6145) (*see* **Note 5**).

7. Primate ES medium (ReproCELL, Cat# RCHEMD001) and 5 ng/mL recombinant human basic fibroblast growth factor (FGF; Wako, Cat# 064-04541) for feeder-dependent human iPS cell maintenance.
8. StemFit AK02N (Ajinomoto) for feeder-free human iPS cell maintenance.
9. iMatrix-511 (Nippi, Cat# 892012).
10. Serum-free differentiation medium for human iPS cells: DMEM/F12 (Thermo Fisher Scientific, Cat# 11320-033) with 2% (v/v) B27 without retinoic acid, 2 mM L-glutamine, 1% (v/v) Insulin-Transferrin-Selenium (Thermo Fisher Scientific, Cat# 41400-045), 1% (v/v) nonessential amino acids, 90 μM 2-mercaptoethanol, and 0.5% (v/v) penicillin/streptomycin.
11. Medium for mouse embryo dissection: DMEM containing 10% FBS.
12. Medium for spinal cord induction: DMEM/F12 containing 10% FBS.
13. Dulbecco's Phosphate-Buffered Saline without calcium or magnesium (PBS(−); Wako, Cat# 045-29795).
14. 0.25% Trypsin-EDTA (Thermo Fisher Scientific, Cat# 25200-056).
15. Cell counter.
16. Dissociation solution (ReproCELL, Cat# RCHETP002).
17. ESGRO Complete Accutase (Merck, Cat# SF006).
18. Type IV collagenase (Thermo Fisher Scientific, Cat# 17104-019) (*see* **Note 6**).
19. Growth factors: Recombinant human activin A (R&D Systems, Cat# 338-AC), recombinant human bone morphogenetic protein 4 (BMP4; R&D Systems, Cat# 314-BP), CHIR99021 (Axon Medchem, Cat# Axon1386), retinoic acid (Sigma-Aldrich, Cat# R2625), Y27632 (Wako, Cat# 257-00511), recombinant human basic FGF (R&D Systems, Cat# 233-FB), recombinant human FGF-9 (R&D Systems, Cat# 273-F9).
20. Accumax cell detachment solution (Merck, Cat# SCR006).
21. Fluorescence-Activated Cell Sorting (FACS) buffer: 1× Hank's Buffered Saline Solution (HBSS; Thermo Fisher Scientific, Cat# 14185-052) with 1% BSA and 4 mM sodium hydrogen carbonate (Wako, Cat# 191-01305).
22. Wash buffer: 1× HBSS with 20% FBS, 4 mM sodium hydrogen carbonate, 1 mM calcium chloride dihydrate (Wako, Cat# 031-00435), and 50 μg/mL DNase I (Worthington, Cat# LS002139).

23. Normal mouse serum control (Thermo Fisher Scientific, Cat# 31881).
24. Antibodies for flow cytometry: biotinylated anti-ITGA8 antibody (R&D Systems, Cat# BAF4076), allophycocyanin (APC)-conjugated anti-mouse CD140a (PDGFRa) antibody (BioLegend, Cat# 135907), phycoerythrin (PE)-conjugated streptavidin (BioLegend, Cat# 405203), PE-conjugated anti-human CD140a antibody (BioLegend, Cat# 323506), and APC-conjugated streptavidin (BioLegend, Cat# 405207).
25. PI buffer: 5 μg/mL propidium iodide (Merck, Cat# 537059) diluted with FACS buffer.
26. Nylon mesh 330 (40-μm pore).
27. Surgical tools: Large surgical scissors, small surgical scissors, microdissecting spring scissors, tissue forceps, microdissecting forceps, razor, and tungsten needle.
28. Dissecting microscope.
29. Sterile disposable 35- and 60-mm petri dishes.
30. Sterile disposable 15-mL polypropylene conical tubes.
31. Sterile disposable 1.5-mL microtubes.
32. Sterile disposable 24- and 96-well plates.
33. Sterile disposable 96-well U-bottom low cell-binding plates (Thermo Fisher Scientific, Cat# 174929).
34. Sterile disposable 96-well V-bottom low cell-binding plates (Sumitomo Bakelite, Cat# MS-9096V).
35. Sterile disposable 5-mL round-bottom polystyrene tubes (BD Falcon, Cat# 352058).
36. Sterile disposable polycarbonate membrane filter, 13 mm, 0.8 μm (Whatman, Cat# 110409).
37. Sterile micropipette.

## 3 Methods

### 3.1 Nephron Induction from Mouse ES Cells

#### 3.1.1 Maintenance Culture of Mouse ES Cells

1. One day before passaging mouse ES cells, seed mitomycin C-treated mouse embryonic fibroblasts (MEFs) at a density of $1.2 \times 10^6$ cells per 0.1% gelatin-coated 60-mm dish.
2. Thaw mouse ES cells in mouse ES cell maintenance medium.
3. Passage mouse ES cells every 2–3 days using 0.25% trypsin-EDTA for dissociation.

*3.1.2 Maintenance in 2i Medium Before Differentiation*

1. Before induction of differentiation, passage mouse ES cells onto feeder-free gelatin-coated dishes in 2i medium ($3 \times 10^5$ cells for 6-well plate or 35-mm dish in 2 mL of medium) (*see* **Notes** 7 and **8**).
2. Culture mouse ES cells in 2i medium for 2 days.

*3.1.3 Differentiation of Mouse ES Cells into NPCs*

(Day 0)

1. Aspirate culture medium, and wash mouse ES cells in 35-mm dish once with 2 mL of PBS(−). Aspirate to remove PBS(−).
2. Add 300 μL of Accutase and incubate for 5 min at 37 °C in a 5% $CO_2$ incubator.
3. Stop the enzymatic reaction by adding 1.7 mL of differentiation medium, and collect the cell suspension into a 15-mL tube using a P1000 micropipette.
4. Take a sample (10 μL) of the cell suspension and count the cells.
5. Centrifuge the remaining cell suspension at 210 × *g* for 4 min, and then discard the supernatant.
6. Add differentiation medium to the cell pellet to obtain a cell density of $2 \times 10^4$ cells/mL (1000 cells/50 μL), and resuspend cells using a P1000 micropipette (*see* **Note 9**).
7. Seed 50 μL of cell suspension into each well of a 96-well U-bottom low cell-binding plate.

(Day 2)

8. After 48 h, use a P200 micropipette to harvest spheres into a 15-mL tube, and wash with 2 mL of PBS(−).
9. Centrifuge spheres at 210 × *g* for 1 min, and then discard the supernatant.
10. Add 500 μL of Accutase and incubate for 5 min in a 37 °C water bath.
11. Stop the enzymatic reaction by adding 1.5 mL of differentiation medium, and pipette up and down five times to dissociate into single cells using a P1000 micropipette.
12. Take a sample (10 μL) of the cell suspension and count the cells.
13. Centrifuge the remaining cell suspension at 210 × *g* for 4 min, and then discard the supernatant.
14. Add differentiation medium containing 1 ng/mL human activin A to obtain a cell density of $1 \times 10^5$ cells/mL ($1 \times 10^4$ cells/100 μL) (*see* **Note 10**).

15. Seed 100 μL of the cell suspension into each well of a 96-well low cell-binding plate.

(Day 3)

16. After 24 h, transfer spheres into another 96-well U-bottom low cell-binding plate. Fill each well with 100 μL of C10 medium containing 10 μM CHIR99021 (*see* **Note 11**).

(Day 4.5)

17. Transfer spheres into 100 μL of C10Y medium containing 10 μM CHIR99021 and 10 μM Y27632 (*see* **Notes 12** and **13**).

(Day 5.5)

18. Transfer spheres into 150 μL of ABC3RY medium containing 10 ng/mL activin A, 3 ng/mL BMP4, 3 μM CHIR99021, 0.1 μM retinoic acid, and 10 μM Y27632 (*see* **Note 14**).

(Day 6.5)

19. Transfer spheres into 200 μL of C1F9Y medium containing 1 μM CHIR99021, 5 ng/mL human FGF9, and 10 μM Y27632.
20. Culture spheres for 2 days to induce NPCs.

#### 3.1.4 Flow Cytometry Analysis of NPCs

(Day 8.5)

1. Collect induced NPC spheres into a 60-mm dish with 4 mL of PBS(−).
2. Transfer spheres into 300 μL of 0.25% Trypsin-EDTA in a 1.5-mL microtube with a minimum amount of PBS(−) (*see* **Notes 15** and **16**).
3. Incubate for 5–6 min in a 37 °C water bath (*see* **Note 16**).
4. Stop the enzymatic reaction by adding 700 μL of DMEM containing 10% FBS and 0.5% DNase I to the cell suspension and pipetting up and down with a P1000 micropipette (*see* **Note 17**).
5. Centrifuge the tube at 210 × *g* for 4 min, and then discard the supernatant.
6. Add 500 μL of wash buffer, and gently pipette to dissociate into single cells using a P1000 micropipette.
7. Centrifuge the tube at 210 × *g* for 4 min, and then discard the supernatant.
8. Add 5–10 μL of normal mouse serum. Tap the tube with your finger to ensure mixing, and place on ice for 10 min (Blocking).
9. During blocking, prepare primary antibody solution: biotinylated anti-ITGA8 antibody (1:40) and APC-conjugated anti-mouse CD140a (Pdgfra) antibody (1:500) diluted in FACS

buffer for mouse NPCs and biotinylated anti-ITGA8 antibody (1:40) and PE-conjugated antihuman CD140a (PDGFRa) antibody (1:50) diluted in FACS buffer for human NPCs (Fig. 2a).

**A**

| Mouse | Primary antibodies | Secondary antibody |
|---|---|---|
| Negative control A | (FACS buffer) | (FACS buffer) |
| Negative control B | (FACS buffer) | PE Streptavidin (1:100) |
| Test sample | Biotinylated anti-ITGA8 antibody (1:40)<br>APC-conjugated anti-mouse CD140a (Pdgfra) antibody (1:500) | PE Streptavidin (1:100) |

| Human | Primary antibodies | Secondary antibody |
|---|---|---|
| Negative control A | (FACS buffer) | (FACS buffer) |
| Negative control B | (FACS buffer) | APC Streptavidin (1:100) |
| Test sample | Biotinylated anti-ITGA8 antibody (1:40)<br>PE-conjugated anti-human CD140a (PDGFRa) antibody (1:50) | APC Streptavidin (1:100) |

**B**

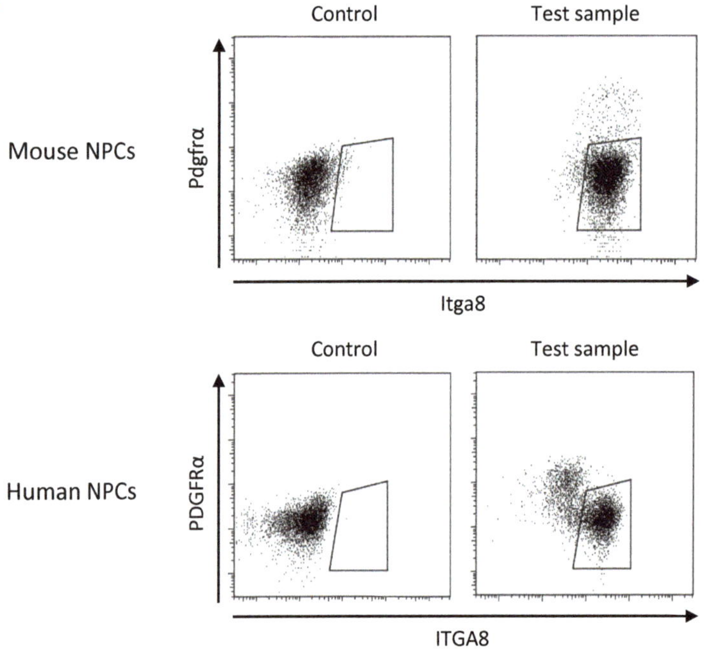


**Fig. 2** Flow cytometry analysis of induced NPCs. (**a**) Conditions of staining to generate quality controls. (**b**) Flow analysis of induced NPCs from mouse ES cells (left) and human iPS cells (right). NPCs are recognized as the ITGA8$^+$/PDGFRa$^-$ population

10. After blocking, aliquot a small amount of cell suspension into two new 1.5-mL microtubes with 100 μL of FACS buffer for use as negative controls A and B (Fig. 2a).
11. Add 100 μL of primary antibody solution to the cell suspension, and react for 30 min on ice (*see* **Note 18**).
12. Wash cells with 300 μL of FACS buffer and tap tube to mix. Centrifuge tube at 210 × *g* for 4 min, and then discard the supernatant.
13. Wash cells again with 400 μL of FACS buffer with gentle pipetting. Centrifuge the tube at 210 × *g* for 4 min, and then discard the supernatant.
14. During washing, prepare secondary antibody solution: PE-conjugated streptavidin (1:100) diluted in FACS buffer for mouse NPCs and APC-conjugated streptavidin (1:100) diluted in FACS buffer for human NPCs (Fig. 2a).
15. Add 100 μL of secondary antibody solution to cell suspension. For negative control A, add 100 μL of FACS buffer. For negative control B, add 100 μL of secondary antibody solution (Fig. 2a). React on ice for 10 min.
16. Repeat **steps 12** and **13**; then proceed to **step 17**.
17. Add 300–500 μL of PI buffer to the cell pellet, and pipette gently to generate a single-cell suspension (*see* **Note 19**).
18. Transfer the cell suspension to a 5-mL round-bottom polystyrene tube by passing through a nylon mesh to remove aggregated cells (*see* **Note 20**).
19. Analyze by flow cytometry (Fig. 2b).

### *3.1.5 Harvesting Mouse Embryonic Spinal Cord*

(Day 8.5)

1. Euthanize a pregnant mouse (E12.5) by cervical dislocation.
2. Wash the abdomen with 70% EtOH and make an incision along the midline. Cut the peritoneum (Fig. 3A-a), and resect the uterus from the abdominal cavity (*see* **Note 21**). Place the uterus in a 60-mm Petri dish with 5 mL of cold DMEM containing 10% FBS.
3. Cut the uterus between each embryos, and dissect the embryos from the uterus (Fig. 3A-b, c).
4. Amputate fetal head and tail from the body, and lay the embryo in a supine position (Fig. 3A-d; *see* **Note 22**).
5. Remove the viscera inside the chest and abdomen, such as the heart, lungs, liver, and intestines (Fig. 3A-e).

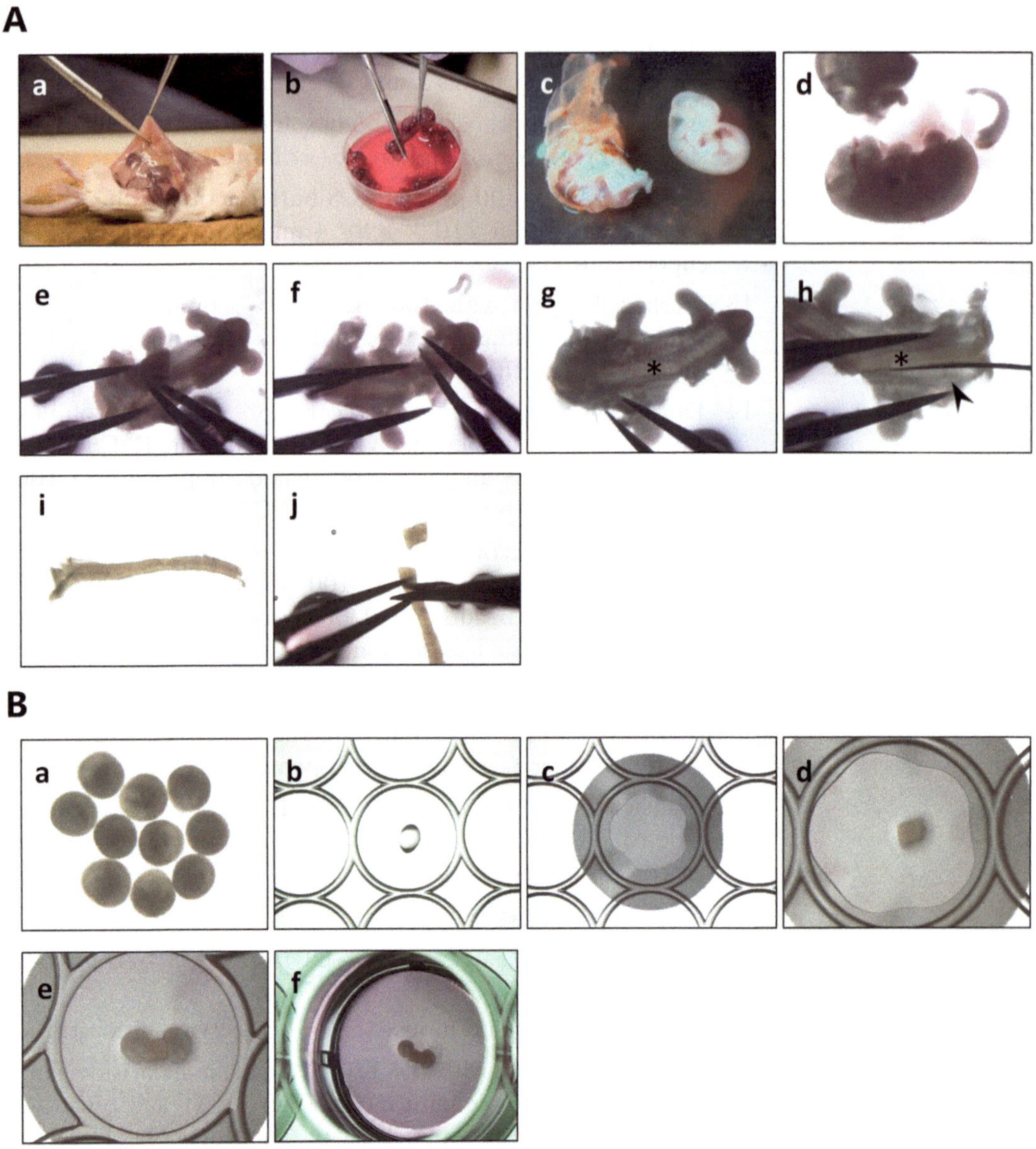


**Fig. 3** Three-dimensional nephron induction from NPCs upon co-culture with mouse embryonic spinal cord. (**a**) Procedure for isolating mouse embryonic spinal cord. An E12.5 embryo was used for dissection in these photos. After making an incision in the peritoneum (**a**), the uterus was dissected (**b**), and the embryos were removed (**c**). The fetal head and tail were amputated (**d**), followed by removal of the viscera, urogenital system, and aorta (**e**, **f**). At this point, the spinal cord was visible (**g**). After resecting somites from the spinal cord (**h**), gross appearance of the naked spinal cord could be observed (**i**). The spinal cord was then cut into small pieces with small scissors (**j**). Asterisks: spinal cord (**g**, **h**). (**b**) Procedure for setting up spinal cord induction. Collect NPC spheres (**a**). Create a small drop of medium (**b**) and place a filter on the drop (**c**). Add a piece of dissected spinal cord (**d**), and place an NPC sphere adjacent to the spinal cord (**e**). Set the filter afloat on the medium (**f**)

6. Remove the urogenital system (mesonephros and metanephros) and aorta (Fig. 3A-f). After removing these tissues, the spinal cord can be observed (Fig. 3A-g).
7. Set the embryo in the prone position. Resect both sides of somites from the spinal cord using a curved tungsten needle (Fig. 3A-h).
8. Resect the notochord from the ventral side of the spinal cord. Transfer the naked spinal cord into a 35-mm Petri dish with 2 mL of cold DMEM containing 10% FBS (Fig. 3A-i).
9. Cut the spinal cord into small pieces using small scissors (Fig. 3A-j).
(Move onto the next step immediately.)

#### 3.1.6 3D-Nephron Induction Using Mouse Embryonic Spinal Cord

1. Pre-warm 1 mL of DMEM containing 10% FBS and 1% penicillin/streptomycin in each well of a 24-well plate at 37 °C in a 5% $CO_2$ incubator.
2. Collect induced NPC spheres into a 60-mm dish with 4 mL of DMEM containing 10% FBS using a cut P20 tip (Fig. 3B-a).
3. Place a small 2-μL drop of DMEM onto the underside of the cover of the 96-well plate (Fig. 3B-b). Place a 0.8-μm polycarbonate filter onto each drop (Fig. 3B-c; *see* **Notes 23** and **24**).
4. Transfer a piece of spinal cord onto the filter using a cut P20 tip (Fig. 3B-d).
5. Place induced NPC spheres adjacent to the cut edge of the spinal cord (Fig. 3B-e). Attach the sphere directly to the spinal cord by aspirating residual medium with a P10 micropipette.
6. Set the filter afloat on the air-fluid interface of the pre-warmed medium (Fig. 3B-f) (*see* **Note 25**).
7. Change half of the medium (0.5 mL) every 3 days.
8. If you have succeeded to induce NPCs, you will observe an epithelializing nephron in some parts of the sphere from 3 days onward and a well-epithelialized sphere from 6 days after induction.

### 3.2 Nephron Induction from Human iPS Cells

#### 3.2.1 Maintenance Culture of Feeder-Dependent Human iPS Cells

1. One day before passaging human iPS cells, seed mitomycin C-treated MEFs at a density of $0.4 \times 10^6$ cells per 0.1% gelatin-coated 60-mm dish.
2. Thaw human iPS cells in Primate ES medium supplemented with 5 ng/mL recombinant human basic FGF and 1% penicillin/streptomycin. Change the medium every 24 h.
3. Passage human iPS cells every 3 days using Dissociation Solution.

*3.2.2 Harvesting Human iPS Cells from MEFs (Feeder-Dependent Culture)*

1. Add type IV collagenase to the culture medium at a final concentration of 1 mg/mL. Wait 1.5–2 h until 70–80% of the clumps have detached from the MEFs (*see* **Note 26**).
2. Detach remaining attached colonies by gently flushing the medium. Harvest colonies into a 15-mL tube.
3. Keep the harvested cell suspension upright and undisturbed to allow cells to collect at the bottom by gravity for 10 min. Remove the supernatant to deplete floating feeder cells.
4. Add 2 mL of PBS(−) and centrifuge at 210 × *g* for 4 min. Discard the supernatant.
5. Add 500 μL of Accutase and incubate cells for 5 min in a 37 °C water bath.
6. Stop the enzymatic reaction by adding 1.5 mL of differentiation medium. Gently pipette several times to dissociate into single cells using a P1000 micropipette.
7. Take a sample (10 μL) of the cell suspension and count the cells.
8. Centrifuge the remaining cell suspension at 210 × *g* for 4 min, and then discard the supernatant. (Move on to Subheading 3.2, **step 5** immediately.)

*3.2.3 Maintenance Culture of Feeder-Free Human iPS Cells*

1. Maintain human iPS cells in StemFit supplemented with 1% penicillin/streptomycin on an iMatrix-511-coated culture dish. Change the medium every 48 h.
2. Passage human iPS cells every 6–8 days ($2.8 \times 10^4$ cells per 60-mm dish in 4 mL of StemFit medium). Add 10 μM of Y27632 to maintenance medium during first 1–2 days after passaging.

*3.2.4 Harvesting Human iPS Cells (Feeder-Free Culture)*

1. Add Y27632 to the culture medium at a final concentration of 10 μM. Incubate the dish at 37 °C in a 5% $CO_2$ incubator for 60 min.
2. Aspirate the medium and wash cultured iPS cells twice with 4 mL of PBS(−). Aspirate to remove the PBS(−).
3. Add 500 μL of Accutase and incubate for 5–10 min at 37 °C in a 5% $CO_2$ incubator (*see* **Note 27**).
4. Aspirate the Accutase.
5. Add 2 mL of PBS(−) (1 mL for a 35-mm dish), and detach iPS cells by gently flushing with PBS(−).
6. Harvest detached cells into a 15-mL tube with 6 mL of maintenance medium for iPS cell culture.
7. Repeat **steps 5** and **6** twice; then proceed to **step 8**.
8. Gently pipette using a 10-mL pipette. Take a sample (10 μL) of the cell suspension and count the cells.

9. Centrifuge the remaining cell suspension at 210 × *g* for 4 min, and then discard the supernatant. (Move on to the next step immediately.)

*3.2.5 Differentiation of Human iPS Cells into NPCs*

(Day 0)

1. Add differentiation medium containing 1 ng/mL activin A, 20 ng/mL human FGF2, and 10 μM Y27632 to obtain a cell density of $1 \times 10^5$ cells/mL ($1 \times 10^4$ cells/100 μL) (*see* **Notes 9** and **28**).
2. Seed 100 μL ($1 \times 10^4$ cells) of cell suspension into each well of a 96-well V-bottom low cell-binding plate.
3. Centrifuge the plate at 210 × *g* for 1 min, and culture at 37 °C in a 5% $CO_2$ incubator.

(Day 1)

4. After 24 h, transfer aggregated spheres to 200 μL of C10 medium containing 10 μM CHIR99021 (*see* **Note 29**).

(Days 3 and 5)

5. Replace half (100 μL) of the medium with C10Y medium containing 10 μM CHIR99021 and 10 μM Y27632 (*see* **Notes 13**, **30–32**).

(Day 7)

6. Transfer spheres to 200 μL of ABC3RY medium containing 10 ng/mL activin A, 3 ng/mL BMP4, 3 μM CHIR99021, 0.1 μM retinoic acid, and 10 μM Y27632 (*see* **Notes 12** and **14**).

(Day 10)

7. Transfer spheres to 200 μL of C1F9Y medium containing 1 μM CHIR99021, 5 ng/mL human FGF9, and 10 μM Y27632.
8. Culture spheres for 3 days to induce NPCs (*see* **Note 33**).

*3.2.6 Dissection of Mouse Embryonic Spinal Cord*

(Day 13)
(*See* Subheading 3.1.5.)

*3.2.7 3D-Nephron Induction Using Mouse Embryonic Spinal Cord*

(Day 13)
(*See* Subheading 3.1.6.)

If you have succeeded to induce NPCs, you will observe an epithelializing nephron in some parts of the sphere from 3 days onward and well-epithelialized spheres from 9 days after induction (Fig. 4a–e).

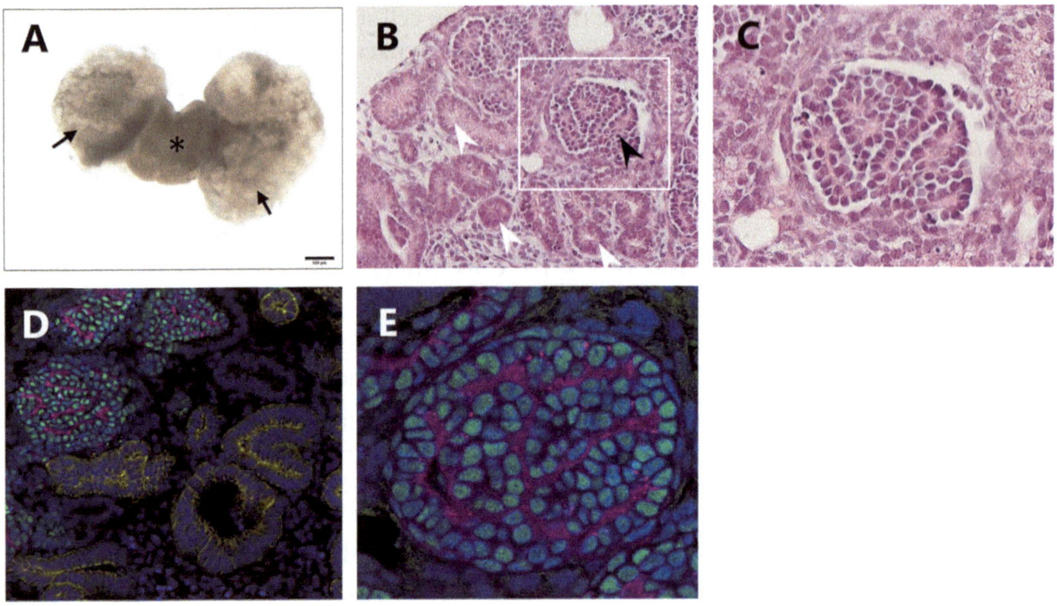


**Fig. 4** Induced nephron tissue from human iPS cells. (**a**) A photograph of induced nephron tissue (black arrows) at day 9 after initiation of co-culture with mouse embryonic spinal cord (asterisk). (**b**, **c**) Hematoxylin and eosin staining of induced nephron tissue at day 9. Higher magnification image (**c**) represents white square in lower magnification image (**b**). Black arrowhead indicates glomerulus; white arrowheads indicate renal tubules. (**d**, **e**) Immunohistochemistry of induced nephron tissue at day 9. Magenta, nephrin (glomerulus marker); green, Wt1 (glomerulus marker); yellow, E-cadherin (distal renal tubule marker)

## 4 Notes

1. Differentiation efficiency may differ between cell lines and maintenance culture conditions, as both affect endogenous signaling levels. Thus, if your cells do not differentiate properly, fine-tune the concentrations of several growth factors as described below.
2. This protocol was originally optimized for the 201B7 cell line. Differentiation efficiency may differ between cell lines and maintenance culture conditions, both of which affect endogenous signaling levels. Although we have confirmed that our protocol works with several human iPS cell lines, if your cells do not differentiate properly, fine-tune the concentrations of several growth factors as described below.
3. We generally use E12.5 or E11.5 pregnant ICR mice to harvest embryonic spinal cords.
4. We use serum-containing 2i medium, unlike commonly used LIF + 2i medium that does not contain serum; this medium is only used to remove feeder cells.

5. IMDM appears to deteriorate over time. Use IMDM within the expiration date, and use mixed differentiation medium within 2 weeks.
6. Dilute type IV collagenase with 1× HBSS at 10 mg/mL.
7. Gelatin coating should be performed for at least 15 min at room temperature.
8. For certain cell lines, this condition may not be suitable to maintain mouse ES cells for long periods of time. Thus, we maintain mouse ES cells on MEFs and transfer to this condition just before differentiation.
9. The initial cell number affects differentiation efficiency. If your cell line does not differentiate properly, decrease or increase the initial cell number.
10. If the differentiation efficiency is not sufficient for your cell line, try to optimize the concentration of activin A (0.5–2 ng/mL) and BMP4 (0–0.3 ng/mL).
11. Transfer spheres using a micropipette and P20 tip with minimal carry over of the previous medium.
12. The size of spheres increases gradually. You should cut the P20 tip to an appropriate size to avoid disrupting the spheres. Transfer spheres to other wells using a micropipette and cut P20 tip.
13. Addition of Y27632 at this step is the result of our induction method development using dissociated mouse embryos.
14. During this culture condition, you may observe dying cells, especially in the periphery of the sphere. However, it is not problematic as long as most cells of the sphere are viable.
15. We generally use up to 15 spheres for 300 μL of 0.25% trypsin-EDTA.
16. In the case of human iPS cell-derived NPCs, we generally use Accumax for dissociation (up to 12 spheres for 500 μL of Accumax). Incubate for 8 min in a 37 °C water bath.
17. In the case of human iPS cell-derived NPCs, we generally do not pipet after incubation.
18. We recommend preparing an appropriate cell density: $1 \times 10^6$ cells per 100 μL of staining solution.
19. The amount of PI buffer should be adjusted for the size of pellet.
20. First, wet the mesh, then turn the mesh upside down, and pass cell suspension through the wet portion of the mesh. This makes it easy to pass the cell suspension through the mesh.
21. When you incise the peritoneum, you should be careful not to injure the intestines.

22. Other mouse embryos should be preserved in cold medium on ice.
23. This drop is used to immobilize the filter.
24. If manipulation of the floating membrane is difficult, you can use Transwell inserts.
25. Take care not to wet the edge of the filter before floating. Otherwise, the filter may sink.
26. Over time, most of the individual colonies begin to float as a clump. Observe every 15 min after 1 h of incubation.
27. Check the dissociation state by pipetting a small amount of Accutase over the colonies. If the colonies are smoothly detached from the dish, you can stop the reaction. If not, continue the incubation for 2-min intervals.
28. If the differentiation efficiency is not sufficient for your cell line, optimize the concentration of activin A (0–3 ng/mL) or BMP4 (0–3 ng/mL).
29. Hereafter, transfer spheres to another well filled with new medium in the 96-well U-bottom low cell-binding plate. Use a micropipette and P20 tip to transfer spheres with minimal carry over of the previous medium.
30. If the differentiation efficiency is not sufficient for your cell line, try to optimize the concentration of BMP4 (0–3 ng/mL) or LDN193189 (1–10 nM).
31. In this step, you do not have to transfer the spheres to new wells. Discard half of the volume of the medium, and add the same volume of fresh medium.
32. Avoid aspirating the spheres or they will break.
33. If the procedure was performed successfully, the induced spheres should appear to be somewhat transparent rather than dark brown at day 13.

## References


1. Constantini F, Kopan R (2010) Patterning a complex organ: branching morphogenesis and nephron segmentation in kidney development. Dev Cell 18(5):698–712
2. Carroll TJ, Park JS, Hayashi S et al (2005) Wnt9b plays a central role in the regulation of mesenchymal to epithelial transitions underlying organogenesis of the mammalian urogenital system. Dev Cell 9(2):283–293
3. Taguchi A, Kaku Y, Ohmori T et al (2014) Redefining the in vivo origin of metanephric nephron progenitors enables generation of complex kidney structures from pluripotent stem cells. Cell Stem Cell 14(1):53–67
4. Taguchi A, Nishinakamura R (2017) Higher-order kidney organogenesis from pluripotent stem cells. Cell Stem Cell 21(6):730–746
5. Sharmin S, Taguchi A, Kaku Y et al (2016) Human induced pluripotent stem cell-derived podocytes mature into vascularized glomeruli upon experimental transplantation. J Am Soc Nephrol 27:1778–1791
6. Kaku Y, Taguchi A, Tanigawa S et al (2017) PAX2 is dispensable for in vitro nephron formation from human induced pluripotent stem cells. Sci Rep 7:4554

# Chapter 9

# Generating Multiple Kidney Progenitors and Cell Types from Human Pluripotent Stem Cells

**Krithika Hariharan, Petra Reinke, and Andreas Kurtz**

## Abstract

Human pluripotent stem cells (hPSCs) have been well known for their ability to generate kidney cell types. We developed a protocol that utilizes a set of growth factors to give rise to kidney progenitors, which when differentiated further in a monolayer gives rise to podocyte precursors, mesangial cells, proximal and distal tubular epithelial cells, and collecting duct cells. This article describes in detail how to obtain each of these segment-specific kidney cell types from hPSCs. Once obtained as a homogenous population, these cells are invaluable for nephrotoxicity testing, for disease modeling, and in tissue engineering approaches such as 3D bioprinting and seeding on acellular matrices and scaffolds.

**Key words** Kidney, Progenitors, iPS, Pluripotent stem cells (PSCs), Directed differentiation, Nephrotoxicity, Kidney development, Tubular epithelial cells, Podocytes, Tissue engineering

## 1 Introduction

The ultimate goal of this study was to derive terminally differentiated kidney cell types. Our protocol is unique in deriving at least five different terminal and two precursor cell types in serum-free, xeno-free conditions and readily transferrable to GMP units and applicable for clinical use from human pluripotent stem cells (hPSCs). Derivation of these cell types opens several topics of concern about their in vitro life span and functionality. For large-scale expansions, it can be postulated that PSC-derived terminal cell types behave similarly to their in vivo counterparts with respect to low mitotic potential, necessitating the usage of larger quantities of PSC as starting material.

The procedure we developed can be divided into three different phases: initial 4 days of metanephric induction, followed by 4 days of mesenchymal epithelial transition of nephron progenitors, and a final 6 days of nephronal cell specification (Fig. 1). The hPSCs are treated with a chemically defined protocol by application of activin A, BMP4, and retinoic acid followed by GDNF, which

Seppo Vainio (ed.), *Kidney Organogenesis: Methods and Protocols*, Methods in Molecular Biology, vol. 1926,
https://doi.org/10.1007/978-1-4939-9021-4_9, 

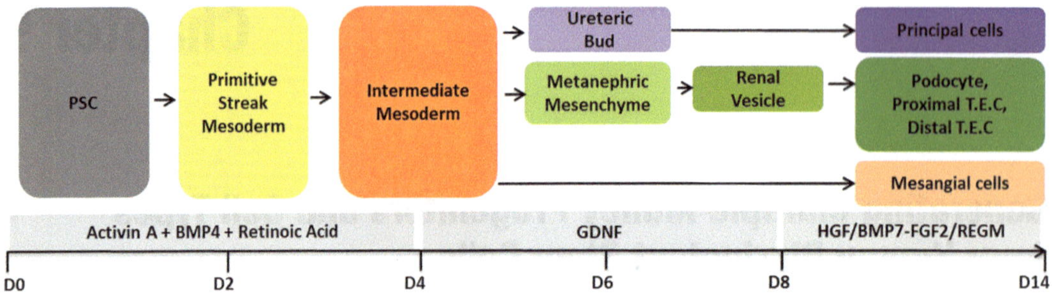


**Fig. 1** Protocol for differentiation of five nephronal cell types via renal vesicle formation

steered hPSCs to the renal lineage-mimicking stages in nephrogenesis and resulted in populations of SIX2$^+$/CITED1$^+$ metanephric mesenchyme (MM)- and of HOXB7$^+$/GRHL2$^+$ ureteric bud (UB)-like cells already by 6 days. Transcriptome analysis corroborated that the PSC-derived cell types at day 8 resemble their renal vesicle and ureteric epithelial counterpart in vivo, forming tubular and glomerular renal cells 6 days later. We demonstrate that starting from hPSC, our in vitro protocol generates a pool of nephrogenic progenitors at the renal vesicle stage, which can be further directed into specialized nephronal cell types including mesangial, proximal tubular, distal tubular, and collecting duct epithelial cells and podocyte precursors after 14 days [1].

Our protocol generates a wide spectrum of building blocks for kidney regeneration from hPSCs. These cells could be used to repair injured kidney segments or used to reconstitute the whole kidney or nephron. Acellular porcine kidney matrices may offer a solution to kidney reconstruction, by providing positional cues that allow functional specification of PSC-derived progenitors. Bioengineering of individual nephrons to the whole kidney may also be achieved by bioprinting the exact segment-specific matrix with segment-specific progenitors. Apart from being used in several combinations with extracellular matrices for tissue reconstitution, the terminal cells may be used for drug screening and toxicity testing. Also, studying molecular mechanisms of genetic disease progression is an opportunity provided by PSCs as they can be derived from patients. Moreover, several key genes leading to fetal death on the occasion of mutations can be knocked down and studied in vitro during the course of development. Thus, PSC-derived kidney progenitors and cells provide a wide range of possibilities to understand and address issues in kidney biology and disease.

## 2 Experimental Design and Materials

### 2.1 hPSC Culture and Maintenance (Subheading 3.1)

1. Human-induced pluripotent stem cell (hiPSC) lines—WISCi004-A (referred to as IMR90-4 iPS, derived from female lung fibroblasts) (WiCell) between 35 and 65 passages, BCRTi004-A [2] and BCRTi005-A [3] derived from urinary cells at passages 25–30 in our lab, and hESC line WAe001-A (WA01) (WiCell) at passages 50–65—have been subjected to our protocol.
2. E8 medium prepared as described in Chen et al. [4].
3. Geltrex LDEV-free, hESC-qualified, Reduced Growth Factor Basement Membrane Matrix (Thermo Fisher Scientific).
4. EDTA—add 500 μL 0.5 M EDTA (pH 8.0) stock into 500 mL calcium−/magnesium-free PBS to obtain 0.5 mM EDTA that can be used for dissociation of hPSC colonies.
5. Cell scraper (TPP 9903) to detach EDTA-treated colonies.
6. KnockOut DMEM/F12 basal medium for washing.
7. Tissue culture dishes.

### 2.2 Differentiation of hPSC to Renal Progenitors (Subheadings 3.2 and 3.3)

1. StemPro Accutase Cell Dissociation Reagent to obtain single cells from hPSC colonies.
2. Y27632 ROCK inhibitor—dissolve the Y27632 in sterile H2O or sterile DMSO with a final concentration of 10 mM (1000×), and then aliquot and store it at −80 °C. The solution is stable for at least 1 year. Use in E8 medium at 10 μM Y27632 (1×) concentration.
3. Differentiation medium used in all stages was prepared using APEL Medium [5] as a base medium, to which growth factors were added as required (Table 1). During stage I, 10 ng/mL activin A, 30 ng/mL rh-bone morphogenetic protein

**Table 1**
**Concentration of growth factors used in this protocol**

| Category | Symbol | Growth factor | Concentration used |
|---|---|---|---|
| Metanephric induction factors | A | Activin A | 10 ng/mL |
| | B4 | BMP4 | 30 ng/mL |
| | RA | Retinoic acid | 1 μM |
| Renal vesicle induction factor | G | GDNF | 150 ng/mL |
| Mesangial factors | B7 | BMP7 | 15 ng/mL |
| | F2 | FGF2 | 10 ng/mL |
| Podocyte factor | HGF | HGF | 50 ng/mL |

4 (BMP4), and 10 μM all-trans-retinoic acid (RA) were added to APEL medium base. Stage II medium is composed of 150 ng/mL glial-derived neurotrophic factor (GDNF; PeproTech) in APEL base. During the course of differentiation, the medium was replenished every 48 h.

#### 2.3 Maturation and Specification of Renal Progenitors to Nephronal Cell Types (Subheading 3.3, step 3)

1. Laminin-521 (BioLamina) in three different mediums.
2. To obtain podocytes, APEL medium with 50 ng/mL rh-hepatocyte growth factor (HGF) is required, and for mesangial cells, APEL medium containing 15 ng/mL rh-bone morphogenetic protein 7 (BMP7) and 10 ng/mL basic fibroblast growth factor (bFGF) is used.
3. Renal epithelial growth medium (REGM; Lonza) is utilized for tubular epithelial differentiation of day 8 cells.

#### 2.4 Endpoint Analysis of hPSC-Derived Renal Cells Using Immune-Fluorescence Staining (Subheading 3.3, step 4)

1. Dulbecco's phosphate-buffered saline (DPBS) with calcium and magnesium ions for washing cells.
2. Fixation buffer—4% paraformaldehyde.
3. Saponin-based/Triton-X-based buffer diluted in water for permeabilization.
4. Donkey/goat serum for blocking buffer depending on secondary antibody species used. DAPI (4',6-diamidino-2-phenylindole) diluted 1:10000 in DPBS. A selection of antibodies that can be used for characterizing the final cell types is presented in Table 2.

## 3 Methods

#### 3.1 hPSCs Preparation and Maintenance

1. Prepare a 1% Geltrex solution and coat a well of a 6-well plate with 1 mL of the solution. Incubate at 37 °C for 30–60 min.
2. Thaw a frozen vial of hPSCs to 37 °C in a water bath until a pellet of ice remains.
3. Transfer the contents of the vial dropwise into 9 mL of E8 medium containing 10 μM Y27632 pre-warmed to room temperature in a 15 mL tube, and centrifuge at 300 × *g* for 5 min.
4. Aspirate the supernatant and resuspend the cell pellet in 2 mL E8 medium with Y27632. Without repeated trituration, seed the cell suspension in the Geltrex-coated well of the 6-well plate after aspirating the Geltrex solution.
5. Culture the cells at 37 °C in a 5% $CO_2$ incubator for 1 day. Aspirate the medium and feed cells with 2 mL of E8 medium to remove Y27632 (*see* **Note 1**).

**Table 2**
**List of antibodies used for immune fluorescence and flow cytometry**

| Name | Species | Description | Company | Catalog number |
|---|---|---|---|---|
| AQP 1 | Rabbit polyclonal | Proximal tubule | Proteintech | 20333-1-AP |
| AQP 2 | Rabbit polyclonal | Collecting duct | Novus Biologicals | NB110-74682 |
| CK8 | Rabbit polyclonal | Whole nephron/ureteric epithelium | Santa Cruz | SC -134484 |
| HOXB7 | Mouse monoclonal | Ureteric bud and tree | R&D Systems | MAB8040 |
| LIM1 | Mouse monoclonal | Intermediate mesoderm/nephric duct | Novus Biologicals | NBP2-01926 |
| Na+/K+-ATPase | Rabbit polyclonal | Proximal tubule | Abcam | AB58475 |
| NKCC2 | Rabbit polyclonal | Loop of Henle | Santa Cruz | SC-133823 |
| PAX2 | Rabbit polyclonal | Intermediate mesoderm/metanephros | Invitrogen | 71–6000 |
| PDGFRß | Rabbit polyclonal | Mesangial cells | Santa Cruz | SC -432 |
| PODXL | Mouse monoclonal | Podocyte membrane | R&D Systems | MAB1658 |
| SIX2 | Mouse monoclonal | Embryonic cap mesenchyme | Abnova (tebu-bio) | H00010736-M01 |
| SLC12A3 | Rabbit polyclonal | Distal tubule and collecting ducts | Novus | NBP1-59699 |
| SYNPO | Mouse monoclonal | Podocyte actin-binding protein | Progen | 65294 |
| WT1 | Rabbit polyclonal | Podocytes, metanephric mesoderm | Santa Cruz | SC-192 |

6. Feed with E8 medium everyday compact stem cell colonies distributed in the well to attain 75–80% confluence (usually within 3–4 days).
7. Use a phase-contrast microscope to observe several compact colonies of cells with a high nuclear to cytoplasm ratio. Cells on the edges of colonies have protrusions that aid in the expansion of the colony.
8. Before passaging, observe the well for spontaneously differentiating cells on the periphery of colonies and the edges of the

well. Using the tip of a micropipette, scrape away the differentiated cells under the microscope at a sterile bench. Remove the scraped cells immediately using KnockOut DMEM/F12 to prevent them from reattachment on the plate.

9. Passage a confluent hPSC culture using the non-enzymatic method [6], by first aspirating the culture medium. Wash the cells with 1 mL of 0.5 mM EDTA to remove all traces of cations that can interfere with the cell detachment process. Add 1 mL of 0.5 mM EDTA and incubate the plate at 37 °C for 3–5 min to allow the colonies to dissociate (*see* **Note 2**).
10. After the stipulated time, aspirate EDTA solution, add 2 mL of E8, detach colonies with a cell scraper, and collect the cell suspension in a 15 mL tube (*see* **Note 3**).
11. Adjust the seeding density for maintenance needs for each individual cell line depending on the proliferation rate and doubling time. Typically in cases of EDTA passage, cells are seeded in 6-well plates at a 1:10 ratio (i.e., contents of each well can be redistributed to ten new wells).

### 3.2 Preparation for Differentiation into Renal Cells

1. hPSC cultures that have attained 80% confluence are eligible to be subjected to our differentiation process, avoiding hassles of replating as single cells (*see* **Note 4**) (proceed to Subheading **3.3**).
2. If the assay in question requires quantification of certain parameters, a standard starting number of hPSCs is preferred for the complete experiment. In this case, a single cell passaging of hPSCs is undertaken using the enzyme Accutase.
3. For single cell passaging, aspirate the spent medium after removing all spontaneously differentiating cells. Wash the cells with DPBS ($Ca^{2+}$- and $Mg^{2+}$-free), and add 1 mL Accutase enzyme to the desired well of a 6-well plate. Incubate at 37 °C for 5–7 min.
4. After 5 min, check every minute to see if all cells are dissociated, and then tap to accelerate the detachment process. Add 2 mL E8 medium containing 10 μM Y27632 to each well in order to dilute Accutase. Collect the contents of the well in a 15 mL tube and centrifuge at 300 × *g* for 5 min.
5. Aspirate the supernatant and resuspend the cell pellet in E8 medium with Y27632. The resulting cell suspension is prepared to arrive at a seeding density of 10,000 cells/$cm^2$ and to be seeded on 1% Geltrex pre-coated plates. After 24 h, remove the medium and feed cells with E8 medium for 2 days (*see* **Note 1**). On the third day, after seeding single cells (day 0), the culture is ready for differentiation.

### 3.3 Differentiation of hPSCs into Renal Lineage

1. **Metanephric induction**—this step results in the specification of hPSCs into intermediate mesodermal cells followed by their differentiation into metanephric mesenchymal (MM) cells and ureteric bud (UB) cells.
   (a) Prepare A.B4.RA medium containing the activin A, BMP4, and retinoic acid in concentrations described in Table 1. Warm freshly prepared medium at room temperature (*see* **Notes 5** and **6**).
   (b) On day 0 of differentiation, aspirate the stem cell maintenance medium, and add 2 mL pre-warmed A.B4.RA medium. Incubate at 37 °C in a 5% $CO_2$ incubator for 48 h.
   (c) On day 2 of differentiation, remove the spent medium and replenish the well with A.B4.RA medium. Considerable cell death and change in morphology of the colonies can be observed. If cell death is greater than 40%, adjust the concentration of retinoic acid in the medium.
   (d) On day 4 of differentiation, observe cultures for the appearance of domes, uniformly distributed in the well (Fig. 2) and can either be harvested at this point or subjected to further differentiation. To obtain more advanced cell types from these progenitors, proceed to Subheading 3.3, **step 2**.
   (e) In our experience, this protocol yields 60–70% SIX2-positive MM cells and 30–40% HOXB7-positive UB cells and 10–15% FOXD1-positive stromal cells by this time point during differentiation. The ratio of cells forming MM-UB-stroma is cell line-specific and has to be determined for every hPSC line.
   (f) Harvest day 4 cells using Accutase enzyme and purify the desired population by flow cytometry and reseed on desired coating for further experiments.

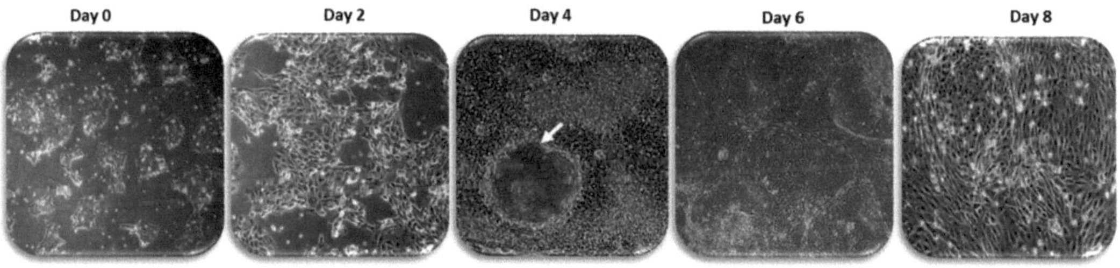


**Fig. 2** Morphological characteristics of PSCs treated with AB4RA-G. PSC colonies at day 0 exposed to A.B4.RA medium form confluent cultures exhibiting dome formation (arrowhead) by day 4. GDNF treatment is followed by rapid proliferation and formation of islands of mesenchymal cells surrounded by epithelial cells at day 8

2. **Generating renal vesicles and ureteric epithelium**—MM cells formed by day 4 when treated further with the nephro-medium undergo mesenchymal epithelial transition to epithelial cells of the renal vesicle (RV); meanwhile, UB progenitor cells also proliferate into ureteric epithelial cells within the next 4 days.
   - (a) To obtain more advanced cell types from day 4 cells, remove the A.B4.RA medium, and add freshly prepared medium containing GDNF (G-medium) as prescribed in Table 1 that is pre-warmed to room temperature.
   - (b) Incubate at 37 °C in a 5% $CO_2$ incubator for 48 h. Replenish G-medium once again and incubate for 48 h.
   - (c) On day 8 of differentiation, a highly confluent culture with several layers of cells accompanied by 3D growth can be visualized under the phase-contrast microscope (Fig. 2).
   - (d) At this stage, the cells stain positively for renal vesicle markers like WT1, JAG1, and LGR5 and ureteric markers like HOXB7 and GRHL2. To obtain the desired compartment-specific precursor, cell sorting can be performed on day 8. If interested in differentiation toward terminal nephron cell types, proceed to Subheading 3.3, **step 3**.
   - (e) As proof of the differentiation, incubate one control well of differentiation for an additional 2–4 days (depending on which cell line) without further medium change. Observe on the last day several round polarized structures with a central lumen and other cord-like structures similar to renal vesicles and ureteric trees growing in the well.
   - (f) Harvest day 8 cells using Accutase enzyme and freeze in medium containing serum and 10% DMSO for future large-scale experiments if cells from the same batch with lesser variations are required.
3. **Maturation and specification of renal progenitors to nephronal cell types**—cells at day 8 are representatives of RV, and ureteric epithelium can be steered further into their terminal identity. The proximal RV population is capable of forming glomerular cells together with the remnant stromal population, whereas the medial and distal RV cells could potentially develop into proximal and distal tubular epithelial cells. Ureteric epithelia mature into collecting duct cells. To obtain a minimum of five identifiable cell types, we use three different protocols.
   - (a) **Deriving cells of the glomerular compartment of the nephron-mesangial-like cells**

(a.1) Prepare rhLaminin-521 solution (10 μg/mL) in DPBS with $Ca^{2+}$ and $Mg^{2+}$, and add 1 mL per well of a 6-well plate. Incubate overnight at 2–8 °C.

(a.2) Aspirate the medium on day 8 cells, wash with DPBS, and add 1 mL of Accutase enzyme for 10 min. From the eighth minute onward, check cells for complete dissociation periodically.

(a.3) Add 2 mL of APEL medium to the cell suspension and collect in a 15 mL tube. Centrifuge at 300 × *g* for 5 min.

(a.4) Aspirate the supernatant and resuspend the pellet in pre-warmed APEL medium containing 15 ng/mL BMP7 and 10 ng/mL FGF2.

(a.5) Seed cells at a density of 15,000 cells/cm$^2$ on a well pre-coated with rhLaminin-521 that has been pre-warmed to room temperature.

(a.6) Incubate at 37 °C in a 5% $CO_2$ incubator and replenish APEL-B7.F2 medium every other day.

(a.7) Culture for 6 days to observe fibroblastic cells that stain positive for markers such as PDGFRb, DESMIN, and ASMA when subjected to immunefluorescence that can be referred to as hPSC-derived mesangial-like cells.

(b) **Deriving cells of the glomerular compartment of the nephron- podocyte-like cells**

(b.1) Prepare a laminin-coated dish and harvest day 8 cells as described in Subheading 3.3, **steps 3a (a.1)–3a(a.3)**.

(b.2) Aspirate the supernatant and resuspend the pellet in pre-warmed APEL medium containing 50 ng/mL HGF.

(b.3) Seed cells at a density of 8000 cells/cm$^2$ on a well pre-coated with rhLaminin-521 that has been pre-warmed to room temperature (*see* **Note 6**).

(b.4) Incubate at 37 °C in a 5% $CO_2$ incubator and replenish APEL-HGF medium every other day.

(b.5) Culture for 6 days to observe huge arborized cells with foot processes and mostly binucleate (Fig. 3) that stain positive for markers such as PODXL, SYNPO, and NPHS2 when subjected to immunefluorescence that can be referred to as hPSC-derived podocyte-like cells.

(c) Deriving cells of the tubular compartment of the nephron and collecting duct.

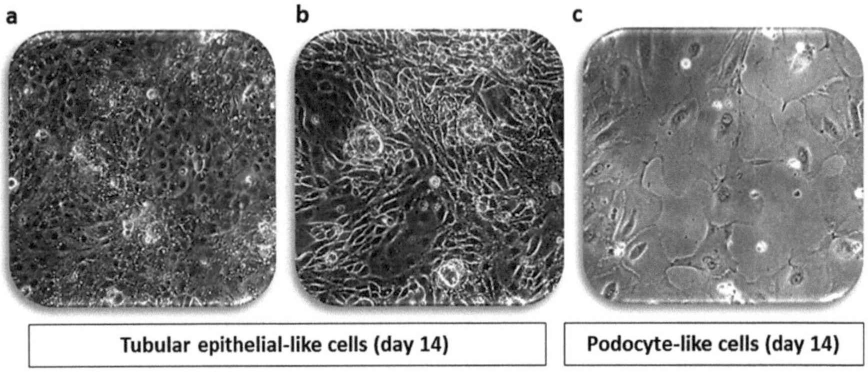


**Fig. 3** Morphology of terminally differentiated cells at day 14. Day 8 cells treated with REGM for 6 days acquire tubular epithelial identity (**a**, **b**). Day 8 cells treated with APEL-HGF acquire podocyte-like morphology (**c**)

(c.1) Prepare a laminin-coated dish and harvest day 8 cells as described in Subheading 3.3, **steps 3a(a.1)–3a (a.3)**.

(c.2) Aspirate the supernatant and resuspend the pellet in pre-warmed renal epithelial growth medium (REGM).

(c.3) Seed cells at a density of 15,000 cells/cm$^2$ on a well pre-coated with rhLaminin-521 that has been brought to room temperature.

(c.4) Incubate at 37 °C in a 5% $CO_2$ incubator and replenish REGM every other day.

(c.5) Culture for 6 days to observe a highly confluent, heterogeneous mass of tubular epithelial cells in several layers (Fig. 3). Closer analysis of the cells at day 14 revealed the presence of several types of hPSC-derived tubular epithelial cells (TECs) that can be sorted by flow cytometry.

(c.6) Observation of tubular epithelial cells:

- Proximal tubular epithelial-like cells—the proximal tubule epithelial cells are characterized by brush border epithelial cells, which are highly absorptive. AQP1+ cells constitute 57% of the tubular epithelial cell mass at day 14.
- Distal tubular epithelial-like cells—sodium chloride cotransporter (NCCT/SLC12A3) and kidney-specific cadherin (CDH16) were expressed in a minor population when subjected to immune-fluorescence, revealing the existence of distal epithelial-like cells.

- Collecting duct-like cells—a distinctive clear zone of cobblestone-shaped aquaporin 2 (AQP2) + cells co-expressing grainyhead-like 2 (GRHL2) and Dolichos biflorus (DB) lectin indicates the presence of collecting duct constituents, i.e., principal-like cells (PC).

(c.7) On day 14, all the terminal cell types can be used for several experiments like seeding acellular matrices, scaffolds, multi-organ chips or for 3D printing and in order to test drugs.

(c.8) Cells typically maintain their morphology for an additional 5 days. Passaging these cells is not recommended due to their highly plastic nature.

4. **Endpoint analysis of hPSC-derived renal progenitors and renal cells**
   (a) Perform the differentiation protocol in parallel on a 96-well plate for immune-fluorescence at day 4, day 8, and day 14 to verify the stage of development attained by the hPSCs in the course of the protocol.
   (b) Aspirate the medium and wash cells with 200 μl DPBS containing $Ca^{2+}$ and $Mg^{2+}$. Add 50 μL of Cytofix™ fixation buffer (BD Biosciences) and incubate for 10 min.
   (c) Aspirate the fixation buffer and wash twice with DPBS.
   (d) For the immune-fluorescent staining of intracellular markers, incubate the cells for at least 15 min, 1× Perm/Wash™ buffer (BD Biosciences) in a buffer containing the cell-permeabilizing agent, saponin.
   (e) Wash with DPBS and add 100 μL of blocking buffer.
   (f) Aspirate the blocking buffer and add 50 μL of the primary antibodies (Table 2) of choice at the recommended dilution. Incubate the plate overnight at 4 °C.
   (g) Wash with DPBS and add 50 μL of secondary antibodies at 1:1000 dilution. Incubate the plate in the dark at room temperature for 60 min.
   (h) Aspirate the secondary antibody and wash twice with DPBS.
   (i) Add DAPI (1:10000) in DPBS and incubate for 10 min in the dark at room temperature.
   (j) Aspirate DAPI and wash cells with PBS. Add 200 μL of DPBS to the cells, and observe the samples under an immune-fluorescence microscope.
   (k) Alternatively, RNA can be harvested at required time points for qPCR analysis of renal markers.

## 4 Notes

1. At Subheadings 3.1, **step 5** and 3.2, **step 5**, if medium containing Y27632 is not withdrawn after 24 h, the cells begin to look spiky and do not begin colony formation.
2. Passaging with EDTA requires a modulation of incubation time such that the fragmentation does not result in single cells.
3. While passaging or thawing, do not triturate the cell suspension several times, as this will result in smaller colonies.
4. Before proceeding with renal differentiation, ensure at least 75–80% confluence of hPSCs as lower cell confluence leads to massive cell death during differentiation. Do not prepare huge stocks of differentiation medium or subject them to temperature fluctuations as the growth factors lose their activity progressively leading to lower efficiency of differentiation.
5. The concentration of retinoic acid is cell line-specific and needs to be titrated for each individual hPSC line. If using the recommended concentration leads to massive cell death, reduce the concentration up to 0.5 μM RA. Retinoic acid has to be freshly thawed before each use.
6. In Subheading 3.3, **step 3a(a.3)**, plating cells at higher densities results in the formation of cord-like structures due to the action of HGF in tubularizing epithelial cells, and hence podocytes will not form.

## Acknowledgments

This work was supported by Berlin-Brandenburg School for Regenerative Therapies and the German Federal Ministry of Education and Research (VIP, FKZ 03V0396, and FKZ 031A303B).

### References

1. Hariharan K, Stachelscheid H, Rossbach B, Oh SJ, Mah N, Schmidt-Ott K, Kurtz A, Reinke P (2018) Parallel generation of easily selectable multiple nephronal cell types from human pluripotent stem cells. Cell Mol Life Sci. https://doi.org/10.1007/s00018-018-2929-2
2. Rossbach B, Hildebrand L, El-Ahmad L, Stachelscheid H, Reinke P, Kurtz A (2016) Generation of a human induced pluripotent stem cell line from urinary cells of a healthy donor using integration free Sendai technology. Stem Cell Res 16(1):133–136
3. Rossbach B, Hildebrand L, El-Ahmad L, Stachelscheid H, Reinke P, Kurtz A (2016) Generation of a human induced pluripotent stem cell line from urinary cells of a healthy donor using an integration free vector. Stem Cell Res 16 (2):314–317
4. Chen G, Gulbranson DR, Hou Z, Bolin JM, Ruotti V, Probasco MD et al (2011) Chemically defined conditions for human iPSC derivation and culture. Nat Methods 8:424–429
5. Ng ES, Davis R, Stanley EG, Elefanty AG (2008) A protocol describing the use of a

recombinant protein-based, animal product-free medium (APEL) for human embryonic stem cell differentiation as spin embryoid bodies. Nat Protoc 3(5):768–776

6. Beers J, Gulbranson DR, George N, Siniscal-chiLI JJ, Thomson JA, Chen G (2012) Passaging and colony expansion of human pluripotent stem cells by enzyme-free dissociation in chemically defined culture conditions. Nat Protoc 7(11):2029–2040

# Chapter 10

# Protocol to Generate Ureteric Bud Structures from Human iPS Cells

Shin-Ichi Mae, Makoto Ryosaka, and Kenji Osafune

## Abstract

The generation of ureteric bud (UB), which is the renal progenitor that gives rise to renal collecting ducts and lower urinary tract, from human-induced pluripotent stem cells (hiPSCs) provides a cell source for studying the development of UB and kidney disease. Here we describe a stepwise and efficient two-dimensional differentiation method of hiPSCs into Wolffian duct (WD) cells. We also describe how to generate three-dimensional WD epithelial structures that can differentiate into UB-like structures.

**Key words** Human pluripotent stem cells, Anterior primitive streak, Anterior intermediate mesoderm, Wolffian duct, Ureteric bud

## 1 Introduction

The reciprocal induction of ureteric bud (UB) and metanephric mesenchyme (MM) facilitates the development of mammalian adult kidney, metanephros. After invading into the surrounding MM, UB undergoes branching morphogenesis and develops the renal collecting duct system and lower urinary tract. MM that condenses beneath UB tips forms the nephron structure following mesenchymal-to-epithelial transition (MET), while MM located above the tips remains undifferentiated. The proximal nephron contains glomerular podocytes and proximal renal tubules, whereas the distal nephron consists of distal renal tubules that fuse with UB [1]. In vitro cultures of these renal progenitor tissues make for good models to study the mechanisms that regulate kidney organogenesis, which are poorly understood.

Human-induced pluripotent stem cells (hiPSCs) are an attractive source for renal progenitor cells because of their developmental potential to differentiate into any cell type in our body [2]. Recent studies have established methods that differentiate hiPSCs to UB structures [3–5]. In this report, we describe a stepwise

Seppo Vainio (ed.), *Kidney Organogenesis: Methods and Protocols*, Methods in Molecular Biology, vol. 1926,
https://doi.org/10.1007/978-1-4939-9021-4_10, 

two-dimensional induction method for WD cells and a simplified procedure that generates UB structures.

## 2 Materials

### 2.1 Reagents

1. Phosphate-buffered saline without calcium or magnesium (PBS (−)).
2. Fetal bovine serum (FBS).
3. StemFit® AK02N (Ajinomoto) (*see* **Note 1**).
4. Essential 6 medium (Thermo Fisher Scientific).
5. iMatrix-511 silk (Nippi).
6. Matrigel matrix growth factor reduced (Matrigel GFR) (Corning).
7. 0.5 M EDTA, pH 8.0.
8. NaCl.
9. 0.5 mM EDTA/PBS solution: To prepare 500 mL of 0.5 mM EDTA/PBS solution, combine 500 mL of PBS (−), 500 μL of UltraPure 0.5 M EDTA, and 0.9 g of NaCl in a sterile 500 mL bottle (*see* **Note 2**).
10. Accutase (Innovative Cell Technologies).
11. Y-27632 (WAKO).
12. Recombinant human/mouse/rat activin A (R&D Systems).
13. CHIR99021 (StemRD).
14. Recombinant human bone morphogenetic protein (BMP)4 (PeproTech).
15. Recombinant human/murine fibroblast growth factor (FGF) 8b (PeproTech).
16. 4-[(E)-2-(5,6,7,8-Tetrahydro-5,5,8,8-tetramethyl-2-naphthalenyl)-1-propenyl]-benzoic acid (TTNPB) (Santa Cruz Biotechnology).
17. A83-01 (WAKO).
18. LDN193189 (Axon Medchem).
19. Recombinant human glial cell derived neurotrophic factor (GDNF) (R&D).
20. Recombinant human FGF acidic (FGF1) (aa 16-155) (R&D).
21. Thiazovivin (Santa Cruz Biotechnology).

### 2.2 Equipment

1. $CO_2$ incubator with control and monitor systems for $CO_2$, humidity, and temperature.
2. Biosafety cabinet for cell culture.

3. Cell culture centrifuge.
4. Cell culture disposables: multi-well tissue culture plates, low-attachment 96-well plates, centrifuge tubes, stripettes, pipettes, and pipette tips.

## 3 Methods

### 3.1 Maintenance of hiPSCs

1. Aspirate the hiPSC maintenance medium (StemFit® AK02N medium) from the culture plates, and wash the cells twice with warm 0.5 mM EDTA/PBS solution.
2. Add warm 0.5 mM EDTA/PBS solution (2 mL/well of 6-well plates) and incubate at 37 °C for 3 min.
3. Aspirate the 0.5 mM EDTA/PBS solution and add StemFit® AK02N medium containing 10 μM Y-27632 (1 mL/well of 6-well plates).
4. Pipet and dissociate the hiPSC colonies into single cells using P1000 pipetman (*see* **Note 3**).
5. Count the cell number.
6. Transfer the hiPSC solution to a 15 mL tube, and add StemFit® AK02N medium containing 10 μM Y-27632 at a density of $2.5 \times 10^4$ cells/mL.
7. Distribute the cells into 6-well plates at a density of $5 \times 10^4$ cells/cm$^2$ (2 mL/well), and add iMatrix-511 silk (0.5 μg/cm$^2$) (*see* **Note 4**).
8. Change the medium every day and grow the cells until the hiPSC colonies reach approximately 70% confluency (*see* **Note 5**).

### 3.2 Anterior Primitive Streak (APS) Induction

1. Repeat Subheading 3.1, **steps 1–5**.
2. Transfer the hiPSC solution to a 15 mL tube, and add StemFit® AK02N medium containing 10 μM Y-27632 at a density of $1.0 \times 10^5$ cells/mL.
3. Distribute the cells into 24-well plates at a density of $2.5 \times 10^4$ cells/cm$^2$ (500 μL/well), and add iMatrix-511 silk (0.5 μg/cm$^2$) (*see* **Note 4**).
4. After overnight incubation at 37 °C, aspirate the medium and wash the cells with warm Essential 6 medium.
5. Add Essential 6 medium containing 100 ng/mL recombinant human/mouse/rat activin A, 3 μM CHIR99021, and 10 ng/mL recombinant human BMP4 (500 μL/well) to the cells, and culture at 37 °C for 1 day.

### 3.3 Anterior Intermediate Mesoderm (IM) Induction

1. After APS induction, aspirate the medium and wash the cells with warm Essential 6 medium.
2. Add Essential 6 medium containing 200 ng/mL recombinant human/murine FGF8, 0.1 μM TTNPB, 10 μM A83-01, and 0.1 μM LDN193189 (1 mL/well) to the cells, and culture at 37 °C for 2 days.
3. Aspirate the medium and wash the cells with PBS (−).
4. Add warm Accutase (300 μL/well) and incubate at 37 °C for 3 min.
5. Pipet and dissociate the cells into single cells using P1000 pipetman.
6. Transfer the cell solution to 5 mL of Essential 6 medium containing 2% FBS, and pipet the cells up and down using P1000 pipetman.
7. Count the cell number.
8. Transfer the cell solution to a 15 mL tube and spin down at 160 × *g* for 5 min.
9. Remove the supernatant and resuspend the cells with Essential 6 medium containing 200 ng/mL recombinant human/murine FGF8, 0.1 μM TTNPB, 1 μM A83-01, 0.1 μM LDN193189, and 10 μM Y-27632 at a density of $2.0 \times 10^5$ cells/mL (*see* **Note 6**).
10. Distribute the cells into Matrigel GFR-coated 24-well plates (*see* **Note 7**) at a density of $1.0 \times 10^5$ cells/cm$^2$ (1 mL/well) and culture at 37 °C for 1 day.

### 3.4 WD Induction

1. After anterior IM induction, aspirate the medium and wash the cells with warm Essential 6 medium.
2. Add Essential 6 medium containing 1 μM CHIR99021, 0.1 μM LDN193189, 200 ng/mL recombinant human/murine FGF8, and recombinant human GDNF (1 mL/well) to the cells, and culture at 37 °C for 2 days.
3. Aspirate the medium and wash the cells with PBS (−).
4. Add warm Accutase (300 μL/well) and incubate at 37 °C for 3 min.
5. Pipet and dissociate the cells into single cells using P1000 pipetman.
6. Transfer the cell solution to 5 mL Essential 6 medium containing 2% FBS, and pipet the cells up and down using P1000 pipetman.
7. Count the cell number.
8. Transfer the cell solution to a 15 mL tube and spin down at 160 × *g* for 5 min.

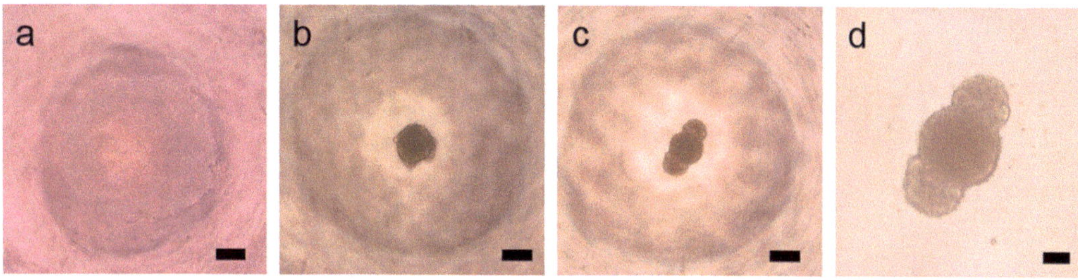


**Fig. 1** Differentiation of WD epithelial aggregates. Morphological changes during the first 2 days of an aggregate formed from dissociated WD cells. Images on days 0 (**a**), 1 (**b**), and 2 (**c**). (**d**) A magnified image of (**c**). Scale bars, 300 μm in (**a–c**) and 100 μm in (**d**)

9. Remove the supernatant and resuspend the cells with Essential 6 medium containing 100 ng/mL recombinant human GDNF, 200 ng/mL recombinant human FGF1, 0.1 μM TTNPB, 3 μM CHIR99021, and 10 μM Thiazovivin at a density of $1.0 \times 10^5$ cells/mL.
10. Distribute the cells into low-attachment 96-well plates at a density of $1.0 \times 10^4$ cells/well (100 μL/well) (*see* **Note 8**) and culture at 37 °C for 2 days (Fig. 1).

### *3.5 UB Induction*

1. After WD induction, aspirate the medium and wash the cells with warm Essential 6 medium.
2. Pipet and detach unwanted cells using P10 pipetman.
3. Aspirate the medium and embed the WD epithelial aggregate in Matrigel (30 μL/well of 96-well plates) (*see* **Note 9**).
4. After the solidification of Matrigel, add Essential 6 medium containing 100 ng/mL recombinant human GDNF, 200 ng/mL recombinant human FGF1, 0.1 μM TTNPB, 3 μM CHIR99021, and 10 μM Thiazovivin (100 μL/well), and culture at 37 °C for 4–15 days (Fig. 2).

## 4 Notes

1. To avoid the degradation of FGF2 activity, StemFit® AK02N medium should be warmed to room temperature without using a water bath.
2. It is important to adjust the osmolarity by using NaCl. The sterilized solution can be stored at 4 °C for up to 6 months.
3. Residual clumps of cells disturb the maintenance of the undifferentiated state.
4. The appropriate dosage of iMatrix-511 silk depends on the cell line. We recommend testing several concentrations of iMatrix-511 silk before final experiments.

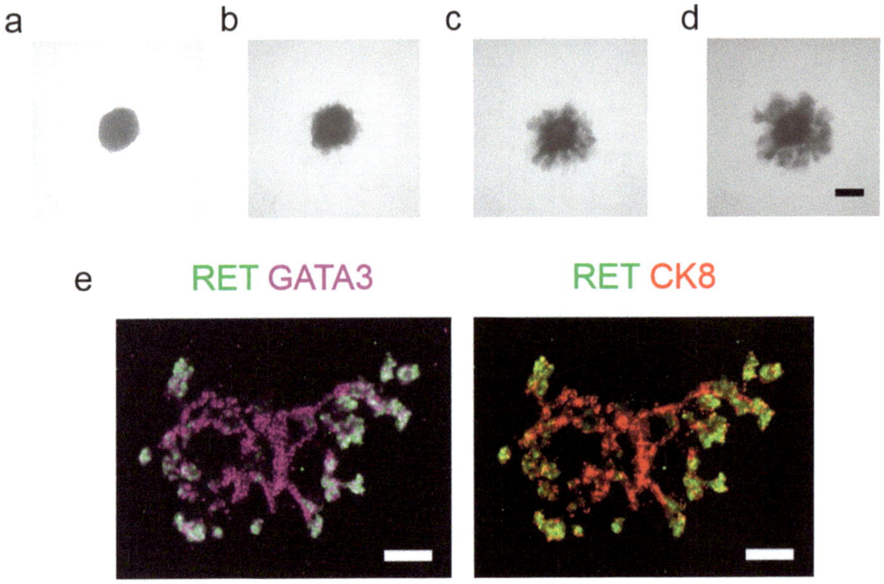


**Fig. 2** Formation of UB structures. Morphological changes during the first 4 days of a WD aggregate embedded in Matrigel. Images on days 0 (**a**), 1 (**b**), 2 (**c**), and 4 (**d**). (**e**) Immunostaining analysis of day 4 structures for a UB tip marker, RET (green), a UB trunk marker, CK8 (red), and a common UB marker GATA3 (purple). Scale bars, 300 μm in (**a–d**) and 100 μm in (**e**)

5. High cell densities cause unwanted differentiation of hiPSCs.
6. Y-27632 is the most commonly used ROCK inhibitor to prevent the apoptosis of dissociated single cells.
7. The appropriate dosage of Matrigel GFR depends on the cell line. The thick coating of Matrigel GFR inhibits the attachment of anterior IM cells.
8. To quickly form cell aggregates, centrifugation (160 × *g* for 2 min) is recommended.
9. Matrigel should be thawed on ice and diluted two times using cold Essential 6 medium. For embedding WD epithelial aggregates, add 30 μL Matrigel solution to each well and incubate at 37 °C for 30 min.

## Acknowledgments

We thank Dr. Peter Karagiannis, CiRA, Kyoto University, for critically reading and revising the manuscript. This study was partially supported by Otsuka Pharmaceutical Co., Ltd.; by the Japan Society for the Promotion of Science (JSPS) through its "Funding Program for World-Leading Innovative R&D on Science and Technology (FIRST Program)" to K.O., Grant-in-Aid for Scientific Research (C) (JSPS KAKENHI Grant Number 24591195) to K.

O., and Grant-in-Aid for Young Scientists (B) (JSPS KAKENHI Grant Number 17K15546) to S.I.M.; and by the Japan Agency for Medical Research and Development (AMED) through its research grant "Projects for Technological Development and Core Center for iPS Cell Research, Research Center Network for Realization of Regenerative Medicine," and "Practical Research Project for Rare/ Intractable Diseases."

## References

1. Costantini F, Kopan R (2010) Patterning a complex organ: branching morphogenesis and nephron segmentation in kidney development. Dev Cell 18:698–712
2. Takahashi K, Tanabe K, Ohnuki M, Narita M, Ichisaka T, Tomoda K, Yamanaka S (2007) Induction of pluripotent stem cells from adult human fibroblasts by defined factors. Cell 131:861–872
3. Xia Y, Nivet E, Sancho-Martinez I, Gallegos T, Suzuki K, Okamura D, Wu MZ, Dubova I, Esteban CR, Montserrat N, Campistol JM, Izpisua Belmonte JC (2013) Directed differentiation of human pluripotent cells to ureteric bud kidney progenitor-like cells. Nat Cell Biol 15:1507–1515
4. Taguchi A, Nishinakamura R (2017) Higher-order kidney organogenesis from pluripotent stem cells. Cell Stem Cell 21:730–746 e736
5. Mae SI, Ryosaka M, Toyoda T, Matsuse K, Oshima Y, Tsujimoto H, Okumura S, Shibasaki A, Osafune K (2018) Generation of branching ureteric bud tissues from human pluripotent stem cells. Biochem Biophys Res Commun 495:954–961

# Part IV

## Protocols to Target the Functional Players in Kidney Ontogenesis

# Chapter 11

# Investigating Aspects of Renal Physiology and Pharmacology in Organ and Organoid Culture

**Melanie L. Lawrence, Mona Elhendawi, and Jamie A. Davies**

## Abstract

Some aspects of renal physiology, in particular transport across tubular epithelia, are highly relevant to pharmacokinetics and to drug toxicity. The use of animals to model human renal physiology is limited, but human-derived renal organoids offer an alternative, relevant system in culture. Here, we explain how the activity of specific transport systems can be assessed in renal organoid and organ culture, using a system illustrated mainly for mouse but that can be extended to human organoids.

**Key words** OCT, OAT, MRP, MATE, Transport, Fluorescent probe

## 1 Introduction

Organ culture of fetal kidney rudiments has traditionally been used to explore renal development [1] and occasionally to explore aspects of viral replication and of nephrotoxicity [2, 3]. Renal physiology has normally been studied in vivo, typically in animals but also to a limited extent in humans, in the context of clinical trials and treatments. The reason for this is obvious: kidneys are usually studied as part of a multi-organ system that controls blood pressure, pH, etc., and much renal function depends on features of the living body not replicated in organ culture, such as flowing blood. The use of animal models as proxies for humans does, however, raise serious problems in fields such as drug development because humans differ from most model animals in their expression and tissue distribution of some transporters and other proteins. For example, rat kidneys express organic cation transporters OCT1 and OCT2; human kidneys express OCT2 but not, or at most very little, OCT1 [4, 5]. These interspecies variations give some drugs very different profiles of efficacy, toxicity, and excretion in model animals compared to humans [6]. The result of this is that more than a fifth of community-acquired acute kidney injury is due

Seppo Vainio (ed.), *Kidney Organogenesis: Methods and Protocols*, Methods in Molecular Biology, vol. 1926,
https://doi.org/10.1007/978-1-4939-9021-4_11, 

to drug toxicity [7] and an estimated 9% of drugs fail clinical trials due to unexpected nephrotoxicity (reviewed in [8]). It is possible that clinical development of a similar number of potentially valuable lead compounds is halted because of nephrotoxicities in animals that would not apply to humans.

For these reasons, there has been a great interest in developing human kidney models from human ES cells or, more commonly, human iPS cells [9]. These can be wild-type cells from healthy donors, cells from a particular type of patient, or mutant cells engineered, for example, using CRISPR technology either to model disease or to investigate hypotheses about human physiology. Currently the most realistic human renal models are organoids that contain multiple cell types, with nephrons, stroma, and collecting tubules (although the macroanatomy is not yet there). There is great hope that these might be useful in drug screening, and some limited data suggests that they are damaged by human nephrotoxicants [10]. There is also hope that they may be useful in helping to analyze why certain classes of small molecules, harmless in animals, are dangerous to human kidneys and even in developing "helper drugs" that protect kidneys against them.

Laboratories developing organoids for use in drug screening or in basic pharmaco-toxicological research need assays to show whether the aspects of tubular physiology relevant to particular pathways of damage are active in their cultures. For the purposes of both toxicology and pharmacology, most renal physiology can be divided crudely into two types: glomerular filtration and transepithelial transport. Glomerular filtration, responsible for some aspects of drug excretion and for some types of toxicity (especially autoimmune and inflammatory), depends on the existence of renal vasculature carrying pressurized blood. These are beyond the capabilities of current in vitro culture systems and so will be ignored here. Transepithelial transport, responsible for recovery of many compounds in filtrate back to the body, responsible for active excretion of organic anions and cations, and partly responsible for pH and ion homeostasis, depends on transporter systems present on tubular epithelia. Its existence does not require flow of blood or urine, although in life transport rates can be extensively modified by these. Transport events are, however, often linked to one another (e.g., potassium influx at the basal side of a cell uses different proteins to its efflux at the apical) and linked to general cell physiology (e.g., many cotransporters move $Na^+$ down-gradient to "pay for" the entropic cost of moving another molecule up-gradient). Mere presence of a transporter protein, detectable immunohistochemically or inferred from a mRNA detected by RT-PCR, is not therefore evidence of transport taking place.

Transport itself can most usefully be studied by using a labeled version of the molecule to be transported. For ions, radio-labels remain the most practical choice, but their use in culture labs is problematic (because of safety gear). In some cases, it may be possible to use microscopic observation of fluorescent reporters of ion concentration, for example, sodium green [11]. Organic molecules, such as glucose, can also be radio-labeled. Fortunately, the tubular transport systems most relevant to drug toxicity are those that bring organic anions or organic cations (including drug metabolites) into and out of cells. These systems are capable of transporting fluorescent organic anions, such as 6-carboxyfluorescein (6-CF, 6-FAM), and fluorescent organic cations, such as 4′,6-diamidino-2-phenylindole (DAPI) [12], so the entry into the cell of these molecules, their accumulation, and/or their exit can be monitored by standard fluorescence live imaging. Note that the common use of DAPI as a nuclear stain depends on cell membranes being disrupted during fixation/postfix processing: uptake is negligible in living cells that do not express the right organic cation transporter systems.

Most transporters involved in transepithelial transport are members of the solute carrier (SLC) protein family, and many had their existence inferred from drug effects long before the actual protein was identified. Some names, for example, the amiloride-sensitive channel (ASC), bear witness to this. Use of the appropriate drug remains the standard method of connecting observed transport of a labeled molecule with a specific cellular mechanism. For this reason, the methods described here use a combination of fluorescent tracers and transport-blocking drugs (Table 1). These drugs can be used to separate questions of uptake and efflux: if a drug is present from the beginning of the tracer experiment, it can test whether uptake seen in the absence of the drug is taking place via the route the drug blocks (Fig. 1a–c). If cells are allowed to take up the tracer unhindered and the drug is then added in medium containing no tracer, the drug can indicate whether cells' loss of the tracer proceeds by the route the drugs block (Fig. 1d). Multiple drugs can be used in parallel to ask more sophisticated questions. Further information about the drugs, including $IC_{50}$, can be found at the Guide to Pharmacology website (the resource is described in [39]).

In setting out the methods below, our intention is not to provide an exhaustive list of all possible tests of transport physiology but rather to illustrate tests of organic anion and cation uptake and efflux through organic anion transporter (OAT) and organic cation transporter (OCT) channels. Readers can then adapt them to their own specific questions. The techniques shown apply to fetal organ cultures and to organoids, but care must be taken when using these techniques on different species (e.g., human vs. mouse) due to interspecies differences in transporter expression and individual transporter substrate specificity. We have described the results of studies of this type in detail elsewhere [40].

**Table 1**
**Channels, fluorescent probes, and pharmacological inhibitors**

| Channel | Fluorescent probe | Inhibitor | Reference for inhibitor |
|---|---|---|---|
| OAT1 (SLC22A6) | 6-CF | Probenecid<br><br>Sensitive to ICA<br>Ellagic acid | Ichida et al. [13]<br>Monien et al. [14]<br>Schulz et al. [15]<br>Whitley et al. [16] |
| OAT2 (SLC22A7) | *6-CF?* | Probenecid (less sensitive to ICA and ellagic acid)<br>Bromosulfophthalein, indomethacin | Enomoto et al. [17]<br><br>Shen et al. [18] |
| OAT3 (SLC22A8) | 6-CF | Probenecid<br>Sensitive to ICA<br>Insensitive to ellagic acid | Hagos et al. [19]<br>Schultz et al. [15]<br>Whitley et al. [16] |
| OAT4 (SLC22A11) | 6-CF | Probenecid | Enomoto et al. [17]<br>Hagos et al. [20] |
| OAT5 (SLC22A19) | *6-CF?* | Probenecid<br>Insensitive to ellagic acid | Youngblood and Sweet (mouse) [21] |
| URAT1 (SLC22A12) | *6-CF?* | Probenecid | Tan et al. [22] |
| MRP1 (ABCC1) | 6-CF | Probenecid<br><br>Verlukast (MK-571) | Gollapudi et al. [23]<br>Bakos et al. [24]<br>Gekeler et al. [25] |
| MRP2 (ABCC2) | 6-CF | Probenecid<br>Verlukast (MK-571) | Bakos et al. [24]<br>Gekeler et al. [25] |
| OCT1 (SLC22A1) | DAPI<br>Rhodamine123 | Clonidine<br>Cimetidine | Zhang et al. [26]<br>Okuda et al. [27] |
| OCT2 (SLC22A2) | ASP+<br>Rhodamine123 | Decynium-22<br>Cimetidine<br>Metformin<br>TPA | Gorboulev et al. [4]<br>Shu et al. [28]<br>Kimura et al. [29]<br>Çetinkaya et al. [30] |
| MATE1 (SLC47A1) | DAPI<br>Rhodamine123 | Cimetidine | Ito et al. [31] |
| MATE2K (SLC47A2) | DAPI<br>Rhodamine123 | Cimetidine | Ito et al. [31] |
| P-gp (MDR1, ABCB1) | Rhodamine123<br>Hoescht 33342 | GF120918<br>Valspodar (PSC-833) | Hyafil et al. [32]<br>Tai [33] |
| BCRP (ABCG2) | Rhodamine123<br>Hoescht 33342 | GF120918<br>Ko143 | Kruijtzer et al. [34]<br>Allen et al. [35] |

Please note that the purpose of this table is to give a positive indication of an active drug: the absence of a drug in a cell in the table does not imply that the drug is known to be inactive on that transporter but rather that there is a better choice of inhibitor. Users of this table are encouraged to view original references for information on concentration, source, etc.
Notes: (i) There is no known rodent ortholog to human OAT4. (ii) Mouse Oat5 (SLC22A19) and human OAT5 (SLC22A10) are not orthologous genes [36]. (iii) Oct1 and Oct2 are both expressed in mouse kidney with Oct1 at higher levels, whereas in humans, OCT1 is mainly expressed in the liver, and OCT2 is the predominant isoform in the kidney. (iv) Mate1 is the only Mate known to be expressed in mouse kidney [37], whereas MATE1 and MATE2K are both expressed in human kidney [38]

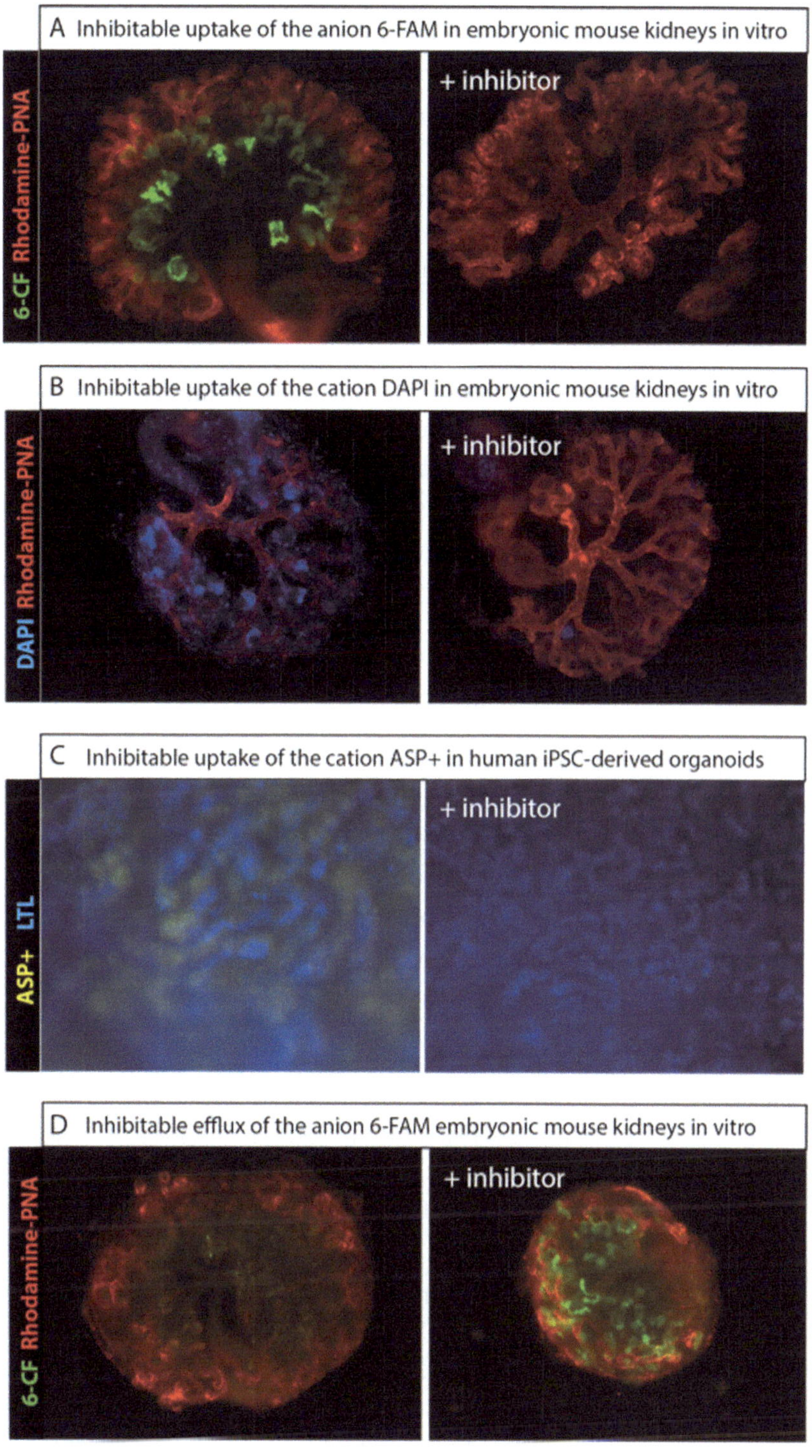


**Fig. 1** Uptake and efflux of fluorescent anions and cations in live, cultured mouse embryonic kidneys or human organoids. (**a**) Uptake of 6-CF in cultured mouse embryonic kidney. Uptake is inhibited when cultured with the anion uptake inhibitor probenecid. (**b**) Uptake of DAPI in cultured mouse embryonic kidney. Uptake is inhibited when cultured with the cation uptake inhibitors cimetidine and metformin. (**c**) Uptake of ASP+ in cultured human renal organoid. Uptake is inhibited when cultured with the cation uptake inhibitor TEA+. (**d**) Efflux of 6-CF in cultured mouse embryonic kidney. Efflux is inhibited, trapping the fluorophore intracellularly, when cultured with the MRP family inhibitor MK-571. Other efflux transporters are blocked throughout the experiment with digoxin

## 2 Materials

Note: This section describes the stock solutions separately, and Subheading 3 describes their dilution into culture medium immediately before use. We have made no studies of the stability of the final dilutions and strongly recommend that the strategy outlined here, of making stable stocks and diluting them into culture medium at the moment of use, is followed.

### *2.1 Live Organic Anion Uptake Assay in Culture*

1. Mouse kidneys or organoids or human organoids in culture (Sebinger culture [41] or grown on filter or membrane; *see* **Notes 2** and **3**).
2. 6-Carboxyfluorescein ('6-CF'/'6-FAM'): 1 mM in distilled, sterile water.
3. Rhodamine-conjugated peanut agglutinin: 5 mg/mL in culture.
4. Probenecid: 250 mM in distilled, sterile water. If the free acid form is purchased, it will need to be dissolved in 500 mM–1 M NaOH, but it is possible and preferable to purchase a water-soluble formulation.
5. Microcentrifuge tubes or falcon tubes (depending on volume required for each treatment).
6. Phosphate-buffered saline (PBS, e.g., Sigma P4417).
7. Culture medium (e.g., Minimal Essential Eagle's Medium, "MEM"; Sigma M5650).
8. Fluorescence microscope.

### *2.2 Live Organic Cation Uptake Assay in Culture (i): DAPI*

1. Mouse kidneys or organoids in culture (Sebinger culture [41] or grown on filter or membrane; *see* **Note 2**).
2. DAPI (4′,6-diamidino-2-phenylindole): 1 mM in distilled, sterile water.
3. Metformin: 100 mM in distilled, sterile water.
4. Cimetidine: 25 mM in distilled, sterile water.
5. Rhodamine-conjugated peanut agglutinin: 5 mg/mL in PBS or medium.
6. Microcentrifuge tubes or falcon tubes (depending on volume required for each treatment).
7. Phosphate-buffered saline (PBS, e.g., Sigma P4417).
9. Culture medium (e.g., Minimal Essential Eagle's Medium, "MEM"; Sigma M5650).
8. Fluorescence microscope.

**2.3 Live Organic Cation Uptake Assay in Culture (ii): ASP(+)**

1. Human renal organoids in culture. We use a published protocol to make renal organoids from hiPSCs [42]. *See* **Note 2**.
2. 4-(4-(Dimethylamino)styryl)-*N*-methylpyridinium iodide (ASP(+)): 10 mM in methanol.
3. Tetrapentylammonium chloride (TPA): 100 mM in sterile distilled water.
4. Culture medium (APEL; STEMCELL Technologies).
5. Phosphate-buffered saline (PBS, e.g., Sigma P4417).
6. Biotinylated lotus tetragonolobus lectin (LTL): 2 mg/mL in sterile water.
7. Streptavidin-conjugated Alexa Fluor 350: 2 mg/mL in sterile PBS (*not* water).
8. Microscope slides.
9. Fluorescence microscope. *See* **Note 6**.

**2.4 Live Anion Efflux Assay in Culture**

1. Mouse kidneys or organoids in culture (Sebinger [41] culture or grown on filter or membrane; *see* **Notes 2**, **3**, and **5**).
2. 6-Carboxyfluorescein (6-CF, 6-FAM): 1 mM in distilled, sterile water.
3. Rhodamine-conjugated peanut agglutinin 5 mg/mL.
4. Digoxin: 10 mM in dimethyl sulfoxide (DMSO).
5. MK-571 (verlukast): 93.5 mM in DMSO.
6. Microcentrifuge tubes or falcon tubes (depending on volume required for each treatment).
7. Phosphate-buffered saline (PBS, e.g., Sigma P4417).
8. Culture medium (e.g., Minimal Essential Eagle's Medium, "MEM"; Sigma M5650).
9. Fluorescence microscope.

## 3 Methods

These methods assume basic competence at renal organ/organoid culture.

**3.1 Live Organic Anion Uptake Assay in Culture**

Organic anions enter the mammalian kidney tubules through specific basolateral and apical transport proteins of the OAT family (part of the SLC22 solute carrier superfamily). This large family of anion transporters can be inhibited from transporting anions with specific compounds. Probenecid is a broad inhibitor of OATs in both human and mouse [14, 17, 19, 21, 22]. The following protocol is suitable for use in mouse fetal kidneys as well as both mouse organoids and human iPS-derived organoids. Rhodamine-conjugated peanut agglutinin is added to each experiment as a method to visualize all the organ/organoid structures.

1. Prepare two tubes of culture medium large enough to hold the volume required for the specific experiment that will submerge the kidney or organoid (e.g., embryonic kidneys or organoids grown in the low-volume Sebinger culture system require 200ul per dish to completely submerge the rudiment, whereas kidneys or organoids grown in a conventional dish will need to be removed from the grid into the dish and will require a larger volume to submerge them in the culture dish).
2. For both tubes, dilute 6-carboxyfluorescein 1/1000 from the 1 mM stock to a final concentration of 1 μM, and dilute rhodamine-conjugated peanut agglutinin 1/500 to a final concentration of 10 μg/mL. This is the "control" preparation.
3. To one of the tubes, dilute probenecid to a final concentration of 2.5 mM. This is the "inhibitor" preparation.
4. The control and inhibitor tubes can at this stage be blind-coded (i.e., labeled in a way that does not indicate which is which, until the code is broken after analysis) by an assistant.
5. All dilutions can be made directly in culture medium (e.g., Dulbecco's Modified Eagle Medium (MEM)) or in PBS.
6. Wash cultured kidneys once in warm or room temperature PBS. This should be sterile if continued culture after the uptake assay is desired.
7. Add enough of the prepared mixtures to the cultured kidneys to submerge them, with some kidneys being treated with the control preparation and some with the inhibitor preparation.
8. Incubate at 37 °C for 60 min.
9. During this time, prepare one tube with probenecid to a final concentration of 10 mM. This should be enough volume to treat all samples including control samples. Dilution can be in culture medium or in PBS.
10. Remove from incubator, and rinse 3× for 1 min each time in warm or room temperature PBS.
11. Submerge kidneys in the 10 mM probenecid and incubate for up to 15 min at 37 °C. This is to block all exits through efflux channels and trap any accumulated fluorophore inside the tubule cells.
12. If using the Sebinger culture method, image immediately directly in the dish using an epifluorescence or confocal microscope (*see* **Note 1**). If grown on membrane or filter, transfer to a glass slide, coverslip, or small culture dish prior to imaging for optimal imaging, ensuring that the live culture remains moist.

### 3.2 Live Organic Cation Uptake Assay in Culture (i)

Organic cations are also taken up in mammalian kidneys through specific transport proteins at the basolateral and apical surfaces of the tubule cells including OCT1 in mouse (SLC22A1), OCT2 in mouse and human (SLC22A2), and the OCTN family of carnitine transporters. The MATE family are H+/cation antiporters with isoforms present in the renal tubules of mouse and human (although there are species differences here as well) and are usually considered to be extrusion proteins in vivo due to the acidic luminal environment. However, MATE1 has been shown to take up DAPI in an $H^+$-independent manner, suggesting that this family is also capable of at least some facilitative transport. The transport activity of OCT1 and OCT2 as well as the MATE transporter family can be competitively inhibited by cimetidine and metformin in mouse and human [29, 43–45]. The following protocol is suitable for use as a functional assay in mouse fetal kidneys and mouse organoids and may be suitable for use in human organoids if the MATE family is expressed. It is important to note that DAPI is transported by the MATE family [46] and OCT1 [12] but is not transported by OCT2. In mouse, both OCT1 and OCT2 transporters are expressed in the kidney, but in humans only OCT2 is expressed in the kidney. MATE1 and MATE2K are also expressed in the human kidney, but expression in human organoids has not been tested. In addition, DAPI is not transported by the OCTN family [46]. Rhodamine-conjugated peanut agglutinin is added to each experiment as a method to visualize all the organ/organoid structures.

1. Prepare two tubes large enough to hold the volume required for the specific experiment that will submerge the kidney (e.g., embryonic kidneys grown in the low-volume Sebinger culture require 200ul per dish to completely submerge the rudiment, whereas kidneys grown in a conventional dish will need to be removed from the grid into the dish and will require a larger volume to submerge them in the culture dish).
2. For both tubes, dilute DAPI 1/1000 to a final concentration of 1 μM and rhodamine-conjugated peanut agglutinin 1/500 to a final concentration of 10 μg/mL, in culture medium or PBS. This is the control preparation.
3. In one of these tubes, add metformin diluted 1/20 to a final concentration of 5 mM and cimetidine to a final concentration of 200 μM. This is the inhibitor preparation.
4. The control and inhibitor tubes can at this stage be blind-coded (i.e., labeled in a way that does not indicate which is which, until the code is broken after analysis) by an assistant.
5. Wash cultured kidneys once in warm or room temperature PBS. This should be sterile if continued culture after the uptake assay is desired.

6. Add enough of the prepared mixtures to the cultured kidneys to submerge them, with some kidneys being treated with the control preparation and some with the inhibitor preparation.
7. Incubate at 37 °C for 30 min.
8. Once DAPI enters the cell, it binds stably to double-stranded DNA; therefore, there is no need to trap the DAPI fluorophore intracellularly, as there was with the anion functional assay.
9. Remove cultures from incubator, and rinse 3× for 5 min each time in warm or room temperature PBS.
10. If using the Sebinger culture method, image immediately directly in the dish using an epifluorescence or confocal microscope (*see* **Notes 1** and **4**). If grown on membrane or filter, transfer to a glass slide, coverslip, or small culture dish prior to imaging for optimal imaging, ensuring that the live culture remains moist.

### 3.3 Live Cation Uptake Assay in Culture (ii)

OCT2 is the main cation transporter expressed at the basolateral membrane of the proximal tubules of the human kidney. While DAPI works as a good fluorescent probe to test the functionality of hMATE1, hMATE2-k, and hOCT1 (mainly expressed in the liver), it has been shown that it is not transported by hOCT2 [46]. 4-(4-(Dimethylamino)styryl)-N-methylpyridinium iodide (ASP(+)) has been used as a fluorescent substrate to test the function of OCT2 [47]. The following protocol is suitable to test the function of OCT2 transporters in human renal organoids. The lectin LTL is added to the organoids to outline the proximal tubules.

1. Incubate the organoids overnight with 20 μg/mL biotinylated LTL in APEL medium.
2. Wash the organoids 3× in room temperature medium for 20 min each.
3. Add Streptavidin Alexa Fluor 350 (diluted to a final concentration of 10 μg/mL in culture medium) to the organoids, and incubate for 1 h at 37 °C.
4. Wash the organoids 3× in culture medium for 20 min each.
5. Transfer the renal organoids into 24-well plate, one organoid per well, and add fresh medium to them (APEL medium, 300 μL/well).
6. For inhibitor treatment, add TPA diluted to a final concentration of 1 mM, in medium, or PBS to the organoids, and incubate for 10 min.
7. Add ASP (final concentration of 10 μM in APEL medium or PBS) (in presence of TPA for inhibitory treatment to the organoid) and incubate for 20 min at 37 °C.

8. Wash twice in ice-cold PBS and image directly in the 24-well culture dish using epifluorescence microscope (*see* **Notes 1** and **6**). Organoids can also be moved to a microscope slide and covered with one drop of PBS to prevent drying out for better imaging.

### 3.4 Live Anion Efflux Assay in Culture

Organic anions are effluxed from the mammalian kidney tubules through specific transport proteins including the apical multidrug resistance protein (MRP) family [48] and, to some extent, the apical breast cancer resistance protein (BCRP) [49]. The activity of these families of anion efflux transporters can be inhibited with specific compounds. The MRP family of extrusion proteins, different isoforms of which are located at the brush border in both human and mouse, can be specifically inhibited with the compound MK-571 (verlukast) [25]. Efflux of anions through basolateral OATP proteins has not been shown, and another apical efflux pump, P-glycoprotein (P-gp, MDR1), preferentially transports hydrophobic or neutral compounds [50]. However, inhibiting these transporters and BCRP with digoxin [51–53] isolates the MRP family of transporters. This specific functional assay for the MRP family is described below. The protocol described can be modified by using careful combinations of inhibitors to assess the live functional activity of other specific efflux pumps or families of extrusion proteins. The following protocol is suitable for use in mouse fetal kidneys and may be suitable for human iPS-derived organoids (although this has not been specifically tested in our group). As with the other functional assays described here, rhodamine-conjugated peanut agglutinin is added to each experiment as a method to visualize all the organ/organoid structures.

1. Prepare three tubes large enough to hold the volume required for the specific experiment that will submerge the kidney or organoid (e.g., embryonic kidneys grown in the low-volume Sebinger culture system require 200 μL per dish to completely submerge the rudiment, whereas kidneys or organoids grown in a conventional dish will need to be removed from the grid into the dish and will require a larger volume to submerge them in the culture dish).
2. All dilutions can be made directly in culture medium (e.g., Dulbecco's Modified Eagle Medium (MEM)) or in PBS.
3. Begin by preloading all cultures (both control and test samples) with fluorophore (6-carboxyfluoreccin). To the first tube, dilute 6-carboxyfluorescein to a final concentration of 1 μM and peanut agglutinin to a final concentration of 10 μg/mL. This is the control treatment.
4. Add enough of the prepared mixture to the cultured kidneys to completely submerge them.

5. Incubate at 37 °C for 60 min.
6. In the second tube, dilute digoxin to a final concentration of 500 μM and vehicle (DMSO) to a final concentration of 0.33% (this is in addition to the digoxin diluted in DMSO). This is the control preparation.
7. In the third tube, dilute digoxin to a final concentration of 500 μM and dilute MK-571 to a final concentration of 30 μM. The total concentration of DMSO for both preparations will be 0.83%. If altering the stock concentrations, ensure that total DMSO concentration does not exceed 1% of the final preparation. This is the inhibitor preparation.
8. After the preloading step, wash cultured kidneys once in warm or room temperature PBS. This should be sterile if continued culture after the uptake assay is desired.
9. *Optional:* Image briefly in the culture dish using an upright epifluorescence microscope to confirm adequate preloading.
10. Add enough of the prepared mixtures to the cultured kidneys to completely submerge them, with some kidneys being treated with the control preparation and some with the inhibitor preparation.
11. Incubate at 37 °C for 30–60 min. This range is to accommodate possible individual differences between ages of fetal embryonic kidneys or organoid transporter expression; fewer efflux channels will result in a longer time being required for measurable efflux to occur.
12. Remove from incubator and rinse 3× for 1 min each in warm or room temperature PBS.
13. If using the Sebinger culture method, image immediately directly in the dish using an epifluorescence or confocal microscope (*see* **Note 1**). If grown on membrane or filter, transfer to a glass slide, coverslip, or small culture dish prior to imaging for optimal imaging, ensuring that the live culture remains moist. Imaging must be carried out immediately once the inhibitors have been washed off as the trapped fluorophore in the MK-571-treated cultures may be lost quickly as it is allowed to efflux out of the cells.

## 4 Notes

1. When imaging, if the cultures are no longer adhering to the glass or filter, remove enough liquid just before imaging to flatten them onto the culture dish surface but still maintaining enough liquid to keep them moist. Do not allow them to dry out.

2. If an inverted microscope is to be used for imaging, grow the cultures in Sebinger culture using large glass coverslips rather than glass slides. If using an upright microscope, any of the culture methods (Sebinger, filter, or membrane) are applicable as long as the objective working depth is acceptable. If grown on a filter or membrane, it may be possible to transfer to a different dish or slide for imaging.
3. 6-Carboxyfluroescein is likely to cause significant background staining on membranes and filters if cultures are grown on these. For this reason, growing on glass may be preferable using the Sebinger culture system.
4. DAPI permeability in live cells is very low, unless the cells express certain transporters. However, DAPI is permeable to apoptotic cells or dead cells with membranes no longer intact. In vitro cultures will always have a certain proportion of apoptotic or dead cells especially in the periphery; DAPI will stain these and may appear to give background; however, the nuclear staining of tubules can be clearly distinguished from the bright apoptotic nuclei or diffuse staining of dead cells.
5. In our assays, we used 20 days or older differentiated human organoids (7 days as monolayer and 13 days as organoids; see published differentiation protocol described in text).
6. ASP(+) has polychromatic emission properties, and therefore care must be taken when choosing fluorescence filters for imaging. The fluorophore can be detected in both red and green channels depending on subcellular localization of the fluorophore [54].

## References


1. Saxen L (1987) Organogenesis of the kidney. Cambridge University Press, Cambridge
2. Saxén L, Vainio T, Toivonen S (1962) Effect of polyoma virus on mouse kidney rudiment in vitro. J Natl Cancer Inst 29:597–631. https://doi.org/10.1093/jnci/29.3.597
3. Saxen L (1983) In vitro model-systems for chemical teratogenesis. In: Kolber A, Wong T, Grant L et al (eds) In vitro toxicity testing of environmental agents, part B. Plenum Press, New York, pp 173–190
4. Gorboulev V, Ulzheimer JC, Akhoundova A et al (1997) Cloning and characterization of two human polyspecific organic cation transporters. DNA Cell Biol 16:871–881
5. Karbach U, Kricke J, Meyer-Wentrup F et al (2000) Localization of organic cation transporters OCT1 and OCT2 in rat kidney. Am J Physiol Ren Physiol 279:F679–F687
6. Knight A (2008) Systematic reviews of animal experiments demonstrate poor contributions toward human healthcare. Rev Recent Clin Trials 3:89–96
7. Loghman-Adham M, Kiu Weber CI, Ciorciaro C et al (2012) Detection and management of nephrotoxicity during drug development. Expert Opin Drug Saf 11:581–596. https://doi.org/10.1517/14740338.2012.691964
8. Benjamin A, da Costa AN, Delaunois A et al (2015) Renal safety pharmacology in drug discovery and development. In: Handbook of experimental pharmacology. Springer, New York, pp 323–352
9. Takahashi K, Yamanaka S (2006) Induction of pluripotent stem cells from mouse embryonic and adult fibroblast cultures by defined factors. Cell 126:663–676. https://doi.org/10.1016/j.cell.2006.07.024

10. Davies J (2014) Engineered renal tissue as a potential platform for pharmacokinetic and nephrotoxicity testing. Drug Discov Today 19:725–729. https://doi.org/10.1016/j.drudis.2013.10.023
11. Amorino GP, Fox MH (1995) Intracellular Na + measurements using sodium green tetraacetate with flow cytometry. Cytometry 21:248–256. https://doi.org/10.1002/cyto.990210305
12. Yasujima T, Ohta K, Inoue K, Yuasa H (2011) Characterization of human OCT1-mediated transport of DAPI as a fluorescent probe substrate. J Pharm Sci 100:4006–4012. https://doi.org/10.1002/jps.22548
13. Ichida K, Hosoyamada M, Kimura H et al (2003) Urate transport via human PAH transporter hOAT1 and its gene structure. Kidney Int 63:143–155. https://doi.org/10.1046/j.1523-1755.2003.00710.x
14. Monien BH, Müller C, Bakhiya N et al (2009) Probenecid, an inhibitor of transmembrane organic anion transporters, alters tissue distribution of DNA adducts in 1-hydroxymethylpyrene-treated rats. Toxicology 262:80–85. https://doi.org/10.1016/j.tox.2009.05.016
15. Schulz K, Hagos Y, Burckhardt G et al (2015) The isoquinolone derived prolyl hydroxylase inhibitor ICA is a potent substrate of the organic anion transporters 1 and 3. Nephron 131:285–289. https://doi.org/10.1159/000442531
16. Whitley AC, Sweet DH, Walle T (2005) The dietary polyphenol ellagic acid is a potent inhibitor of hOAT1. Drug Metab Dispos 33:1097–1100. https://doi.org/10.1124/dmd.105.004275
17. Enomoto A, Takeda M, Shimoda M et al (2002) Interaction of human organic anion transporters 2 and 4 with organic anion transport inhibitors. J Pharmacol Exp Ther 301:797–802
18. Shen H, Liu T, Morse BL et al (2015) Characterization of organic anion transporter 2 (SLC22A7): a highly efficient transporter for creatinine and species-dependent renal tubular expression. Drug Metab Dispos 43:984–993. https://doi.org/10.1124/dmd.114.062364
19. Hagos FT, Daood MJ, Ocque JA et al (2017) Probenecid, an organic anion transporter 1 and 3 inhibitor, increases plasma and brain exposure of N-acetylcysteine. Xenobiotica 47:346–353. https://doi.org/10.1080/00498254.2016.1187777
20. Hagos Y, Stein D, Ugele B et al (2007) Human renal organic anion transporter 4 operates as an asymmetric urate transporter. J Am Soc Nephrol 18:430–439. https://doi.org/10.1681/ASN.2006040415
21. Youngblood GL, Sweet DH (2004) Identification and functional assessment of the novel murine organic anion transporter Oat5 (Slc22a19) expressed in kidney. Am J Physiol Renal Physiol 287:F236–F244. https://doi.org/10.1152/ajprenal.00012.2004
22. Tan PK, Ostertag TM, Miner JN (2016) Mechanism of high affinity inhibition of the human urate transporter URAT1. Sci Rep 6:34995. https://doi.org/10.1038/srep34995
23. Gollapudi S, Kim C, Tran B et al (1997) Probenecid reverses multidrug resistance in mutidrug resistance-associated prtoein-overexpressing HL60/AR and H69/AR cells but not in P-glycoprotein-overexpressing HL60/Tax and P388/ADR cells. Cancer Chemother Pharmacol 40:150–158
24. Bakos E, Evers R, Sinkó E et al (2000) Interactions of the human multidrug resistance proteins MRP1 and MRP2 with organic anions. Mol Pharmacol 57:760–768
25. Gekeler V, Ise W, Sanders KH et al (1995) The leukotriene LTD4 receptor antagonist MK571 specifically modulates MRP associated multidrug resistance. Biochem Biophys Res Commun 208:345–352. https://doi.org/10.1006/bbrc.1995.1344
26. Zhang L, Schaner ME, Giacomini KM (1998) Functional characterization of an organic cation transporter (hOCT1) in a transiently transfected human cell line (HeLa). J Pharmacol Exp Ther 286:354–361
27. Okuda M, Saito H, Urakami Y et al (1996) cDNA cloning and functional expression of a novel rat kidney organic cation transporter, OCT2. Biochem Biophys Res Commun 224:500–507
28. Shu Y, Bello CL, Mangravite LM et al (2001) Functional characteristics and steroid hormone-mediated regulation of an organic cation transporter in Madin-Darby canine kidney cells. J Pharmacol Exp Ther 299:392–398
29. Kimura N, Masuda S, Tanihara Y et al (2005) Metformin is a superior substrate for renal organic cation transporter OCT2 rather than hepatic OCT1. Drug Metab Pharmacokinet 20:379–386
30. Çetinkaya I, Ciarimboli G, Yalçinkaya G et al (2003) Regulation of human organic cation transporter hOCT2 by PKA, PI3K, and calmodulin-dependent kinases. Am J Physiol Renal Physiol 284:F293–F302. https://doi.org/10.1152/AJPRENAL.00251.2002

31. Ito S, Kusuhara H, Yokochi M et al (2012) Competitive inhibition of the luminal efflux by multidrug and toxin extrusions, but not basolateral uptake by organic cation transporter 2, is the likely mechanism underlying the pharmacokinetic drug-drug interactions caused by cimetidine in the kidney. J Pharmacol Exp Ther 340:393–403. https://doi.org/10.1124/jpet.111.184986
32. Hyafil F, Vergely C, Du Vignaud P, Grand-Perret T (1993) In vitro and in vivo reversal of multidrug resistance by GF120918, an acridonecarboxamide derivative. Cancer Res 53:4595–4602
33. Tai HL (2000) Technology evaluation: Valspodar, Novartis AG. Curr Opin Mol Ther 2:459–467
34. Kruijtzer CMF, Beijnen JH, Rosing H et al (2002) Increased oral bioavailability of topotecan in combination with the breast cancer resistance protein and P-glycoprotein inhibitor GF120918. J Clin Oncol 20:2943–2950. https://doi.org/10.1200/JCO.2002.12.116
35. Allen JD, van Loevezijn A, Lakhai JM et al (2002) Potent and specific inhibition of the breast cancer resistance protein multidrug transporter in vitro and in mouse intestine by a novel analogue of fumitremorgin C. Mol Cancer Ther 1:417–425
36. Wu W, Baker ME, Eraly SA et al (2009) Analysis of a large cluster of SLC22 transporter genes, including novel USTs, reveals species-specific amplification of subsets of family members. Physiol Genomics 38:116–124. https://doi.org/10.1152/physiolgenomics.90309.2008
37. Lickteig AJ, Cheng X, Augustine LM et al (2008) Tissue distribution, ontogeny and induction of the transporters Multidrug and toxin extrusion (MATE) 1 and MATE2 mRNA expression levels in mice. Life Sci 83:59–64. https://doi.org/10.1016/j.lfs.2008.05.004
38. Masuda S, Terada T, Yonezawa A et al (2006) Identification and functional characterization of a new human kidney-specific H+/organic cation antiporter, kidney-specific multidrug and toxin extrusion 2. J Am Soc Nephrol 17:2127–2135. https://doi.org/10.1681/ASN.2006030205
39. Harding SD, Sharman JL, Faccenda E et al (2018) The IUPHAR/BPS Guide to PHARMACOLOGY in 2018: updates and expansion to encompass the new guide to IMMUNOPHARMACOLOGY. Nucleic Acids Res. https://doi.org/10.1093/nar/gkx1121
40. Lawrence ML, Chang C-H, Davies JA (2015) Transport of organic anions and cations in murine embryonic kidney development and in serially-reaggregated engineered kidneys. Sci Rep 5:9092. https://doi.org/10.1038/srep09092
41. Sebinger DDR, Unbekandt M, Ganeva VV et al (2010) A novel, low-volume method for organ culture of embryonic kidneys that allows development of cortico-medullary anatomical organization. PLoS One 5:e10550. https://doi.org/10.1371/journal.pone.0010550
42. Takasato M, Er PX, Chiu HS et al (2015) Kidney organoids from human iPS cells contain multiple lineages and model human nephrogenesis. Nature 526:564–568. https://doi.org/10.1038/nature15695
43. Tsuda M, Terada T, Mizuno T et al (2009) Targeted disruption of the multidrug and toxin extrusion 1 (Mate1) gene in mice reduces renal secretion of metformin. Mol Pharmacol 75:1280–1286. https://doi.org/10.1124/mol.109.056242
44. Matsushima S, Maeda K, Inoue K et al (2009) The inhibition of human multidrug and toxin extrusion 1 is involved in the drug-drug interaction caused by cimetidine. Drug Metab Dispos 37:555–559. https://doi.org/10.1124/dmd.108.023911
45. Ciarimboli G, Deuster D, Knief A et al (2010) Organic cation transporter 2 mediates cisplatin-induced oto- and nephrotoxicity and is a target for protective interventions. Am J Pathol 176:1169–1180. https://doi.org/10.2353/ajpath.2010.090610
46. Yasujima T, Ohta K-y, Inoue K et al (2010) Evaluation of 4′,6-diamidino-2-phenylindole as a fluorescent probe substrate for rapid assays of the functionality of human multidrug and toxin extrusion proteins. Drug Metab Dispos 38:715–721. https://doi.org/10.1124/dmd.109.030221
47. Wilmer MJ, Saleem MA, Masereeuw R et al (2010) Novel conditionally immortalized human proximal tubule cell line expressing functional influx and efflux transporters. Cell Tissue Res 339:449–457. https://doi.org/10.1007/s00441-009-0882-y
48. Kruh GD, Belinsky MG (2003) The MRP family of drug efflux pumps. Oncogene 22:7537–7552. https://doi.org/10.1038/sj.onc.1206953
49. Allikmets R, Schriml LM, Hutchinson A et al (1998) A human placenta-specific ATP-binding cassette gene (ABCP) on chromosome 4q22 that is involved in multidrug resistance. Cancer Res 58:5337–5339
50. Wang RB, Kuo CL, Lien LL, Lien EJ (2003) Structure-activity relationship: analyses of

p-glycoprotein substrates and inhibitors. J Clin Pharm Ther 28:203–228. https://doi.org/10.1046/j.1365-2710.2003.00487.x

51. Mikkaichi T, Suzuki T, Onogawa T et al (2004) Isolation and characterization of a digoxin transporter and its rat homologue expressed in the kidney. Proc Natl Acad Sci U S A 101:3569–3574. https://doi.org/10.1073/pnas.0304987101

52. Hughes J, Crowe A (2010) Inhibition of P-glycoprotein-mediated efflux of digoxin and its metabolites by macrolide antibiotics. J Pharmacol Sci 113:315–324

53. Pavek P, Merino G, Wagenaar E et al (2005) Human breast cancer resistance protein: interactions with steroid drugs, hormones, the dietary carcinogen 2-amino-1-methyl-6-phenylimidazo(4,5-b)pyridine, and transport of cimetidine. J Pharmacol Exp Ther 312:144–152. https://doi.org/10.1124/jpet.104.073916

54. Morozova GI, Dobretsov GE, Dubur GI et al (1981) 4-(n-Dimethylaminostyryl)-1-methylpyridinium fluorescence in a living cell. Tsitologiia 23:916–923

# Chapter 12

# Virus as Renal Functional Genetic Tools

Abhishek Sharma and Sandhanakrishnan Cattavarayane


## Abstract

Viral vectors enable efficient transfection of ectopic DNA into hard to transfect cells. Viral vectors are normally used to obtain permanent modification of target cells, and tissues expect for the cases where integrase-deficient viruses are used. Here we describe a method to stably transfect metanephric mesenchyme cells isolated from the murine embryonic kidney at day E11.5. Using this method, it is possible to transfect hard to transfect cells and successfully evade host tissue immune response. Due to these advantages, this method has become one of the most frequently used in generating stable cell line, manipulation of tissues, and gene therapy.


**Key words** Viral vectors, Gene transfer, Lentivirus, Metanephric mesenchyme

## 1 Introduction

Viruses depend on their host cell machinery to complete its life cycle. The virus has to bind to the cell membrane of the host and transfer its genetic material into the cytoplasm. The viral genome is transcribed and replicated by the host cell machinery. Different types of viruses including retrovirus, adenovirus, adeno-associated virus, and herpes simplex virus are used in the laboratory for applications like gene therapy [1].

Transfection of primary cell lines using methods like electroporation, nucleofection, lipofection, and calcium phosphate transfection is inefficient and leads to cell death due to toxicity [2]. The mouse embryonic kidney like any other primary tissue is hard to transfect with ectopic DNA. Efficient delivery of nucleic acid into these hard to transfect primary cells is achieved by means of viral vectors [3]. Viral vectors come with different advantages like high efficiency, immune evasion, stable expression, etc.

Most commonly used γ-retroviral vectors have a limited host range and can infect only dividing cells [4], while lentiviruses, a subclass of retroviruses, can infect both dividing and nondividing cells [5, 6]. But both γ-retroviruses and lentiviruses have a

Seppo Vainio (ed.), *Kidney Organogenesis: Methods and Protocols*, Methods in Molecular Biology, vol. 1926,
https://doi.org/10.1007/978-1-4939-9021-4_12, 

limitation in the size of gene the vector can host [7]. Adenoviruses encounter an active immune response and hence are not suitable to use in tissue context [8].

## 2 Materials

Virus production and transduction take place in a biosafety level 2 laboratory. The experiments were performed with the laboratory personnel wearing proper protection from gloves and lab coats. Viral particles produced were tested for replication incompetence. It was ensured that the transduced cells were devoid of any live virus before removing them outside of the virus laboratory.

We used lentiviral vectors to transduce disassociated metanephric mesenchymal (MM) cells. Third-generation lentiviral vectors were used to produce viruses containing green fluorescent protein (GFP) reporter genes (Fig. 1). This system uses a four-plasmid system containing a packaging vector pLenti-GFP, three helper plasmids pMD2.G, and pMDL g/pRRE and pRSV-Rev.

1. Human embryonic kidney 293T (HEK-293T) cells.
2. Lipofectamine 2000.
3. Opti-MEM.
4. Growth medium—DMEM, low glucose, 10% FCS, 0.1% penicillin/streptomycin.
5. 0.44 μm syringe filter.

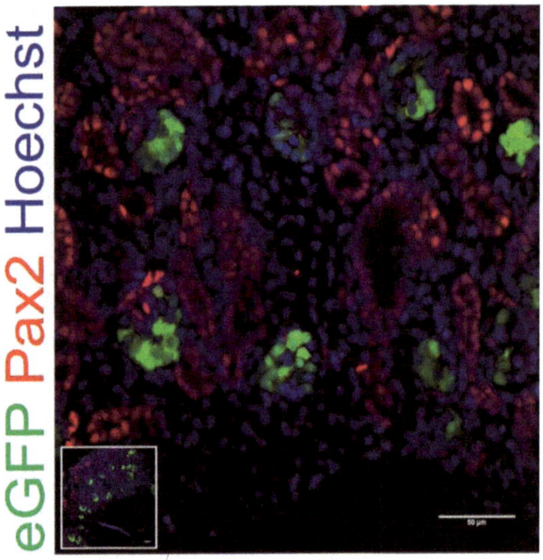


**Fig. 1** Virus-mediated knockdown of Lim-1 (Lim1 shRNA-eGFP) in dissociated metanephric mesenchyme. Epithelial structures are stained with Pax2 (Junttila et al. unpublished data)

### 2.1 Dissociation of MM

#### 2.1.1 Pancreatin-Trypsin Solution and Collagenase

1. Tyrode's solution—137 mM NaCl, 2.7 mM KCl, 1 mM $MgCl_2$, 1.8 mM $CaCl_2$, 0.2 mM $Na_2HPO_4$, 12 mM $NaHCO_3$, and 5.5 mM D-glucose.
2. Pancreatic-trypsin solution—0.25 g of pancreatin from porcine pancreas (Sigma) and 0.17 g NaCl were dissolved into 20 mL of sterile water on a magnetic stirrer for 3–4 h at room temperature. The solution was left on the magnetic stirrer at 4 °C overnight. The solution was centrifuged at 2700 × *g* for 10 min; the supernatant was aliquoted and stored at −20 °C. 0.45 g of trypsin from porcine pancreas (Sigma) was dissolved in 2 mL of pancreatin solution (see above), and adjust the final volume to 20 mL with Tyrode's solution on ice. The pH of the final solution was adjusted between 7.2 and 7.8. Pancreatin-trypsin solution was aliquoted and stored at −20 °C.
3. Earle's balanced salt solution—1.8 mM $CaCl_2$, 5.3 mM KCl, 0.8 mM $MgSO_4$, 117 mM NaCl, 26 mM $NaHCO_3$, 1 mM $NaH_2PO_4.H_2O$, and 5.6 mM glucose.
4. Collagenase III (Worthington)—Collagenase solution was prepared by diluting it in Earle's balanced salt solution to obtain the final concentration of 9400 U/mL. The collagenase solution was aliquoted and stored at −20 °C.

#### 2.1.2 Preparation and Dissection of Embryos

1. Dumont #5 forceps, fine tip.
2. Surgical scissors.
3. 20-gauge hypodermic needles.
4. 5 mL syringes.
5. Zoom stereomicroscope with transmitted light.
6. 1× Dulbecco's PBS without calcium and magnesium.
7. 10 cm glass and plastic petri dishes.
8. Pancreatin-trypsin solution (*see* Subheading 2.1.1).
9. Collagenase III (9400 U/mL).
10. DNase I (1 U/mL) (Thermo Scientific).
11. Growth medium: Dulbecco's modified Eagle's medium (DMEM) with GlutaMAX-I and 1000 mg/mL glucose (Gibco), 10% FBS, 100 U/mL penicillin, 0.1 mg/mL streptomycin, and 10 mM HEPES.

### 2.2 Transduction of MM

1. Biosafety hood.
2. 4 mg/mL polybrene (Sigma).

## 3 Methods

All the methods described below were carried out inside a biological safety hood.

### 3.1 Lentivirus Production

1. Nearly confluent (70–80%) HEK 293T cells were grown on Corning CellBIND 10 cm cell culture dish.
2. One hour prior to transfection, the media in the cell culture dish was replaced with 5 mL of fresh pre-warmed growth medium.
3. To prepare the transfection mix, 80 μL of Lipofectamine 2000 was added to 1 mL of Opti-MEM and incubated at room temperature for 5 min.
4. In another tube, 20 μg of DNA was added in the following proportion. pLenti-GFP (target vector): pVSVG/pMDL g/pRRE/pRSV-Rev was used in 3:1:1:1 to 1 mL of Opti-MEM.
5. The DNA solution was mixed with Opti-MEM containing Lipofectamine 2000 drop by drop with gentle tapping of the tube.
6. The above mixture was incubated at room temperature for 20 min.
7. The mixture was then added on top of the cells and swirled gently, and cells were placed into humidified incubator (+37 °C, 5% $CO_2$) (*see* **Note 3**).
8. 24 h post transfection, the medium containing the transfection mix was removed carefully, and 5 mL of fresh medium was added.
9. The medium supernatant containing the viral particles was collected every 12 h for 3–4 days and stored at +4 °C.
10. The viral supernatants were pooled and centrifuged at 180 × *g* for 5 min and filtered through a 0.44 μm filter.
11. The filtered viral supernatant was then concentrated by ultracentrifugation at 100,000 × *g* for 2 h.
12. The supernatant was discarded, and the viral pellet resuspended in complete Dulbecco's phosphate buffered saline (1/100th of original volume).
13. The resuspended virus is aliquoted and frozen at −80 °C until further use (*see* **Note 3**).

### 3.2 Dissociation of MM

1. Mouse embryo 11.5 days post coitum was obtained; at this time the ureteric bud will have invaded into MM.
2. After dissecting out the uterus, it was cut open in Dulbecco's PBS (without calcium and magnesium) using scissors and forceps.

3. The embryos were carefully transferred to a clean 10 cm plate filled with Dulbecco's PBS (without calcium and magnesium).
4. The embryos were removed from the yolk sac under stereomicroscope using needles attached to syringes.
5. The embryo was cut into half between cranial and the hind limbs and the tail was shortened.
6. The caudal part of the embryo was placed with the ventral side toward the bottom, and the hind limbs were excised.
7. The remaining caudal embryo was placed on its side, and the dorsal part was cut off just ventral to the spinal cord.
8. Any excess tissue around the exposed metanephroi was removed.
9. Pancreatin-trypsin solution was melted in a low-binding 1.5 mL Eppendorf tube.
10. In a 6 cm glass petri dish, warm culture medium was added.
11. The collected metanephroi was added to the pancreatin-trypsin solution for 30 s and immediately transferred to the petri dish containing warm culture medium.
12. Any remnants of the Wolffian duct were removed. By scratching the mesenchyme gently with one needle, the ureteric bud was exposed and pulled out while holding the tissue in place with another needle.
13. The collected MM was added to the dissociation solution prepared by adding 40 μL of collagenase and 10 μL of DNase to 280 μL of physiological buffer.
14. The cell suspension was incubated at 37 °C for 20 min in an atmosphere of 5% $CO_2$.
15. The MM is dissociated into a single-cell suspension by pipetting carefully under a stereomicroscope.
16. 600 μL of warm culture medium was added, and the cells were pelleted by centrifuging at 1380 × *g* for 5 min at room temperature.
17. The supernatant was discarded, and warm culture medium was added along with 50 ng/mL of BMP7 and 100 ng/mL FGF2, and the cells were kept on ice.

### 3.3 Lentivirus Transduction

1. Under the biosafety hood, polybrene was added to the metanephric mesenchyme cells to make a final concentration of 4 μg/mL.
2. The cells were divided into two low-bind 1.5 mL Eppendorf tubes carefully marking them.
3. In a fresh tube, warm culture medium containing BMP7 and FGF2 was added followed by virus solution. The final volume

was adjusted so that the growth factor concentration remains 50 ng/mL and 100 ng/mL for BMP7 and FGF2, respectively.

4. The transfection mix was incubated at 37 °C in 5% $CO_2$ for 2 h.
5. Subsequently, the cells were pelleted at 1380 × *g* for 5 min at room temperature.
6. The supernatant was removed and discarded. Warm culture medium containing BIO (GSK-3 inhibitor) was added.
7. The cells were pelleted at 1380 × *g* for 20 min, and the pellet was incubated at 37 °C in 5% $CO_2$.
8. After 16 h of induction with BIO (GSK-3 inhibitor), the pellet was carefully removed using a glass pipette and carefully placed into Trowell culture.

## 4 Notes

1. HEK 293T cells detach very easily post transfection. Care should be taken while collecting virus and adding new medium. The medium should be gently added dropwise to the sides of the cell culture dish. Care should be taken while removing and putting back the dishes from and into the incubator. The number of attached cells will directly affect the final viral titer (Unpublished observation).
2. The pellet can be immediately resuspended or can be placed at +4 °C overnight to improve the efficiency of resuspension. Overnight protocol helps in better resuspension and thereby improves the efficiency of infection (Unpublished observation).
3. Once the virus is concentrated and resuspended, it should be aliquoted and frozen. Lentiviral particles pseudotyped with VSVG are quite strong and can withstand multiple freeze-thaw cycles. But the efficiency of infection decreases with every round of freeze-thaw. It is better to have aliquots of desired size, which can be completely utilized for infection upon thawing.

### References


1. Bouard D, Alazard-Dany D, Cosset FL (2009) Viral vectors: from virology to transgene expression. Br J Pharmacol 157(2):153–165. https://doi.org/10.1038/bjp.2008.349
2. Maurisse R, De Semir D, Emamekhoo H et al (2010) Comparative transfection of DNA into primary and transformed mammalian cells from different lineages. BMC Biotechnol 10:9. https://doi.org/10.1186/1472-6750-10-9
3. Junttila S, Saarela U, Halt K et al (2015) Functional genetic targeting of embryonic kidney progenitor cells ex vivo. J Am Soc Nephrol 26 (5):1126–1137. https://doi.org/10.1681/ASN.2013060584
4. Spiegel M, Bitzer M, Schenk A et al (1998) Pseudotype formation of Moloney murine leukemia virus with Sendai virus glycoprotein F. J Virol 72(6):5296–5302

5. Naldini L, Blomer U, Gallay P et al (1996) In vivo gene delivery and stable transduction of nondividing cells by a lentiviral vector. Science 272(5259):263–267
6. Yamashita M, Emerman M (2006) Retroviral infection of non-dividing cells: old and new perspectives. Virology 344(1):88–93
7. Yi Y, Noh MJ, Lee KH (2011) Current advances in retroviral gene therapy. Curr Gene Ther 11 (3):218–228
8. Muruve DA (2004) The innate immune response to adenovirus vectors. Hum Gene Ther 15(12):1157–1166. https://doi.org/10.1089/hum.2004.15.1157

# Chapter 13

# Gene Editing in 3D Cultured Nephron Progenitor Cell Lines


## Zhongwei Li, Toshikazu Araoka, and Juan Carlos Izpisua Belmonte



## Abstract

During kidney development, Six2+ nephron progenitor cells (NPCs) generate nephrons, the functional units of the kidney. Isolating and expanding NPCs in vitro have enabled wide applications in both basic and translational research. Here we describe the methods to derive NPC lines from mouse embryonic kidneys in a 3D culture format as well as gene editing in the NPC lines.



**Key words** Nephron progenitor cell (NPC), 3D culture, NPSR culture medium, Gene editing, CRISPR-Cas9


## 1 Introduction

End-stage kidney disease (ESRD) patients must receive renal replacement therapies like hemodialysis or renal transplantation to survive. Nephron progenitor cells (NPCs) give rise to all nephrons, the functional units of the kidney [1]. Unlike some fish species where NPCs exist throughout their lifetime and can produce new nephrons upon kidney injury [2], mammalian NPCs are transient, amplifying cells during development [3] and are not yet identified in the adult. Thus, to use it for alternative strategies to treat kidney failure is urgently needed, and one approach that holds great promise is the derivation of authentic nephron progenitor cell lines in vitro with intact generative capacity.

Kidney development has been extensively studied in mice, and various niche signals for the proper specification, proliferation, and differentiation of murine nephron progenitors have been identified [4–6]. This accumulated knowledge has greatly facilitated efforts in extending the life span of primary murine NPCs in monolayer culture for up to several weeks [7–11]. In contrast to the 2D monolayer culture, we observed that in a 3D culture format with proper NPSR (nephron progenitor self-renewal) medium, NPCs

Zhongwei Li and Toshikazu Araoka contributed equally to this work.


Seppo Vainio (ed.), *Kidney Organogenesis: Methods and Protocols*, Methods in Molecular Biology, vol. 1926,
https://doi.org/10.1007/978-1-4939-9021-4_13,

can be expanded long-term (more than 1 year) with high purity and without losing their nephrogenic potential [12].

Here we describe in detail the 3D culture system that enables the stable long-term in vitro propagation of functional murine nephron progenitors using NPSR medium and the gene editing in these NPC lines. Long-term cultured NPCs can be harnessed for rapid and efficient generation of nephron organoids, thereby providing an accessible system for modeling kidney development, renal toxicity testing, gene editing, and disease modeling.

## 2 Materials

### *2.1 NPSR Medium (see also Notes 1–4)*

Basal medium: DMEM/F12 (1:1) (1×), Invitrogen, Cat. No. 11330-032.

Supplements:

| Reagent name | Company | Cat. no. | Final concentration |
|---|---|---|---|
| GlutaMAX-I (100×) | Invitrogen | 35050-079 | 1× |
| MEM NEAA (100×) | Invitrogen | 11140-050 | 1× |
| 2-Mercaptoethanol (55 mM) | Invitrogen | 21985-023 | 0.1 μM |
| Pen Strep (100×) | Invitrogen | 15140-122 | 1× |
| B-27 Supplement (50×), minus vitamin A | Invitrogen | 12587-010 | 1× |
| ITS Liquid Media Supplement (100×) | Sigma | I3146-5ML | 1× |
| BMP-7 | R&D | 354-BP-010 | 50 ng/mL |
| FGF2 | Peprotech | 100-18B | 200 ng/mL |
| Heparin | Sigma | H3149-100KU | 1 μg/mL |
| Y-27632 | Enzo | ALX-270-333-M025 | 10 μM |
| Mouse LIF | Millipore | ESG1107 | 1000 units/mL |
| CHIR99021 | Reagents Direct | 27-H76 | 1 μM |
| LDN193189 (optional) | Reagents Direct | 36-F52 | 5 to 50 nM |
| A83-01 (optional) | Stemgent | 04-0014 | 0.05 to 0.5 μM |

### 2.2 Dissection Medium

Basal medium: DMEM (1×), Invitrogen, Cat. No. 11995-040.
Supplements:

| Reagent name | Company | Cat. no. | Final concentration |
|---|---|---|---|
| GlutaMAX-I (100×) | Invitrogen | 35050-079 | 1× |
| MEM NEAA (100×) | Invitrogen | 11140-050 | 1× |
| 2-Mercaptoethanol (55 mM) | Invitrogen | 21985-023 | 0.1 μM |
| Pen Strep (100×) | Invitrogen | 15140-122 | 1× |
| FBS | Gemini | 100-106 | 10% |

### 2.3 Other Materials

TrypLE (Invitrogen, Cat. No. 12563-011).
40 μm cell strainer (Greiner Bio-One, Cat. No. 542040).
U-bottom low-attachment 96-well plates (Thermo, Cat. No. 174929).
Accumax (Innovative Cell Technologies, Cat. No. AM-105).
DNeasy Blood & Tissue Kits (Qiagen, 69506).
pGEM®-T Easy Vector Systems (Promega, A1360).

## 3 Methods

### 3.1 Derivation of Mouse NPC Lines from Any Mouse Strain

Previously we described the derivation of mouse NPC lines from *Six2*-GFP knock-in mouse strain by first using FACS to purify the SIX2+ NPCs and the derivation of NPC lines from any mouse strain by first using "culture-dependent purification" (CDP) method to purify the NPCs on laminin-coated plates [12]. Here, we describe an improved, much easier method to derive mouse NPC lines from any mouse strain by utilizing the NPSR medium-based selective expansion of NPCs over other embryonic kidney cell types.

1. To get timed embryonic kidneys from given mouse strain, male and female mice with desired genotypes are put into the same cage in the evening, and plugs are checked the following morning. Midday of plug positive is designated as day 0.5. Pregnant mice are sacrificed at embryonic days from E11.5 to E12.5. The embryonic kidneys are dissected out from mouse embryos with standard dissection technique [13] and are immediately placed in dissection medium on ice before dissociation.
2. Collect kidneys into 1.5 mL tubes with dissection medium.
3. Carefully aspirate dissection medium and wash the kidneys once with sterile PBS.

4. Add 300 μL of TrypLE, and incubate in 37-degree incubator for 15–20 min to dissociate the embryonic kidneys.
5. Add 500 μL dissection medium to the tube, pipette up and down gently 20 to 30 times to further dissociate the embryonic kidneys.
6. Filter the cell suspension through 40 μm cell strainer to remove the remaining cell clumps.
7. Spin down the cells at 300 g for 5 min, and then resuspend the cells in NPSR medium, and seed into U-bottom low-attachment 96-well plates at ~3000 cells/100 μL per well.
8. Spin down the 96-well plate at 300 g for 3 min to make sure the cells form a monolayer at the bottom of the plate. Place the plate into 37-degree incubator, and the 3D aggregate will be formed following overnight incubation. Change the medium every 2 days (each well with 100 μL fresh NPSR medium), and after 4–5 days' growth, the aggregate reached 0.5–1 mm in diameter with 50,000 to 100,000 cells in cell number. At this time point, they are ready for passage.
9. Follow the following protocol in Subheading 3.2 to passage the aggregates. The percentage of NPCs in the whole cell population in the aggregate is increased gradually after each passage. After three passages, more than 95% in the aggregate will be NPCs. At this time point, stable NPC line is successfully derived (*see* also **Note 1**).

### 3.2 Passage of Mouse NPC Lines

1. Collect 3D aggregates to 1.5 mL tubes. Note: Since the aggregates are very soft in nature (mesenchymal origin), do not use 200 or 20 μL pipette tips to transfer the aggregates because the aggregates will be disrupted. 1 mL pipette tips are routinely used in our lab for this purpose.
2. Aspirate the culture medium and wash once with 1 mL sterile PBS.
3. Aspirate PBS, and add room temperature Accumax to the NPC aggregates (the volume of Accumax depends on the number of NPC aggregates to be dissociated). Our experience is 1–2 aggregates, 50 μL; 3–5 aggregates, 100 μL; and 10–20 aggregates, 200–300 μL.
4. Incubate NPCs with Accumax in 37-degree incubator for 7–10 min.
5. Add 500 μL dissection medium, and pipette up and down very gently for no more than five times. The clumps of cells should be invisible after 2–3 times pipetting. Never use force for pipetting the NPCs to obtain single cells, which will negatively affect their self-renewal thereafter.
6. Spin down the cells at 300 g for 5 min, and resuspend the cell pellet with NPSR medium (*see* **Notes 2–4**).

7. Split the NPCs at the ratio around 1:30 to U-bottom low--attachment 96-well plates.
8. Spin down the 96-well plates at 300 g for 3 min, and then place the plates back to 37-degree incubator.
9. Change the medium with 100 μL fresh NPSR medium every 2 days, and passage the NPC aggregates every 4–5 days after the aggregate reached the optimal size for passaging (0.5–1 mm in diameter). In our hand, 1:30 every 4 days is routinely used.
10. Repeat the passage cycles for maintenance and expansion of mouse NPC lines (*see* also **Notes 5–7**).

### 3.3 Clonal Expansion of Mouse NPCs with Helper NPCs

NPCs cultured in NPSR/3D condition cannot be cultured as single cells with currently unknown reasons. To circumvent this problem to enable its applications such as gene editing, we developed this method. The following protocol details an example to generate mCherry+/puromycin-resistant (mCherry/puro for short) single cell clones.

First, a mixture of mCherry/puro NPCs is generated using lentivirus infection:

1. Lentivirus that expresses both mCherry and puromycin-resistant gene was generated via a previously published protocol [14]. One aggregate of NPCs at around 100,000 cells is dissociated and passaged according to protocol in Subheading 3.2 of this chapter, at the split ratio of 1:5 into 5 wells of low-attachment 96-well plate. Only 50 μL NPSR medium, but not 100 μL, was used for each well. 1 μL of concentrated mCherry/puro lentivirus particles is added to each well, together with polybrene at final concentration of 10 μg/mL. After mixing, the plate was spun down at 300 g for 3 min and then placed into 37-degree incubator. After 10 to 12 h, the aggregates are formed. Aggregates are transferred into a 1.5 mL Eppendorf tube and washed with 1 mL PBS for three times. After wash aggregates are placed into new low-attachment wells with fresh NPSR medium (100 μL per well). Note: the volume of lentivirus particles here is just an example. Different virus particle generated with different method has quite different titers and thus the infection volume should be empirically tested.
2. Two days later, change the medium with 100 μL fresh NPSR medium; 3 days later, change the medium to NPSR supplemented with 0.5 μg/mL puromycin. Culture the aggregate with NPSR+puromycin for two passages to eliminate the cells without lentivirus integration. After two passages, the aggregate has bright and homogeneous mCherry expression under the fluorescence microscope (more than 99% mCherry+).

Next, single clones are generated with helper NPCs:

3. Dissociate mCherry/puro mixture NPCs into single cells, calculate the cell densities, and dilute it to 0.5 cell/10 μL in NPSR medium.
4. Add 10 μL dissociated mCherry/puro cells at 0.5 cell/10 μL to each well of low-attachment plate. Spin down the plate at 300 g for 5 min, and then label the wells that have only one cell in each well. According to our experience, serial dilution might be the best choice since FACS-based single cell sorting causes a lot of damage to NPCs. On average, 30–70% of the wells should be single cells as long as the cell densities are calculated accurately.
5. Dissociate wild-type NPCs and dilute it to 300 cells/10 μL. Add 10 μL of wild-type NPCs to the wells with single mCherry/puro cell.
6. Spin down at 300 g for 5 min and place the plate back to 37-degree incubator.
7. Add 30 μL NPSR medium to each well after 2 days; add another 50 μL NPSR medium to each well after 4 days.
8. Change the medium with 100 μL NPSR medium after 6 days, and change the medium every 2 days for another 4 to 6 days until the aggregates reach the size suitable for passage. At this time point, change to NPSR medium containing 0.5 μg/mL puromycin to kill the helper wild-type NPCs.
9. Keep changing medium with NPSR+puromycin until the aggregates grow big enough for passaging. After puromycin selection the whole aggregate represents a clonally expanded mCherry/puro NPC line.

### 3.4 Gene Editing in Mouse NPC Lines

Routine gene manipulations such as gene overexpression and knockdown can be achieved in cultured NPC lines either transiently or stably following standard transfection or lentiviral infection, respectively. The NPC lines are also amenable to gene editing. Here we use CRISPR-Cas9-based gene *Nphs1* knockout as an example.

1. For the lentiviral-based CRISPR vector, the Cas9 vector used was modified from the lenti-CRISPR vector (Addgene, plasmid #49535) with the replacement of original EFS to CMV promoter in order to enhance the expression of Cas9 in target cells and cloning individual gRNAs into the vector. The gRNA sequence targeting mouse *Nphs1* is "*gctgctggtgatgggagcta*." The production, purification, and titration of lenti-CRISPR virus were described previously [14].

2. The produced lenti-CRISPR virus is used to infect NPC lines following protocols described in Subheading 3.3 of this chapter.
3. Three days after lentiviral infection, 0.5 μg/mL of puromycin is used to select for CRISPR-Cas9-integrated NPCs. As described in Subheading 3.3 of this chapter, to select for single cell clones of Nphsl KO NPCs, puromycin-selected mixed population of CRISPR-Cas9-integrated NPCs is dissociated into single cells, and one cell is mixed with 300–500 helper NPCs without expression of puromycin-resistant gene to form aggregate in NPSR medium. Another 7 to 10 days is allowed for the aggregate to grow, and puromycin is then used to selectively kill the helper NPCs.
4. Genomic DNA of established single cell clones is then extracted using Qiagen DNeasy Blood & Tissue Kit following manufacturer's instructions. The sequences surrounding the gRNA targeted site are amplified by PCR. DNA sequencing is used to verify if mutations are generated.
5. To further discriminate between biallelic and monoallelic mutations, TA cloning with pGEM-T Easy Vector System is carried out following manufacturer's instructions using PCR product flanking the targeted site. Plasmids from single *E. coli* clones each carrying one copy of the PCR product are sent for sequencing. Analysis is then carried out by comparing the sequencing results with PubMed reference sequences of *Nphs1* (*see* also **Note 8**).

## 4 Notes

1. When using the described method to derive NPC lines from unpurified embryonic kidney cells, it is normal to see the formation of tubular structures in the first two passages since there are tubular cells such as UB cells in the cell mixture. The tubular structures are difficult to dissociate, and they can be removed during passaging. After three passages, NPCs will dominate the cell population in the aggregates.
2. FGF2 is unstable in 37 degree, and thus if the medium is warmed up using 37-degree water bath every time before use, the activity of FGF2 will be lost fairly quickly. Loss of FGF2 activity will lead to dramatic cell death of NPCs in culture, similar to what was observed in the removal of FGF2 from the culture medium. We do not recommend pre-warm-up for the NPSR medium before use. Keeping the medium in the hood at room temperature for 5–10 min is enough. Apart from the potential variations of these two components, the

other components are very stable even after 1-month daily use in our hand.

3. We noticed that B27 supplement in the NPSR medium varies from batch to batch in supporting NPC growth, similar to what was observed in neural stem cell culture. We usually test a few different lots and stock those batches with good activities to support NPC growth.
4. As optional components for the NPSR medium, A83-01 and LDN193189 allow mouse NPCs to grow at seeding numbers as low as 50 to 100 cells, which is not allowed in NPSR medium without the two components (minimal 300 cells in our hand). The addition of A83-01 and LDN193189 also has dramatic beneficial effects in supporting NPC growth even when "bad" batches of B27 supplements are used. Optimal concentration of A83-01 and LDN193189 needs to be tested experimentally on different NPC lines. We recommend the ranges for A83-01 from 0.05 to 0.5 μM and for LDN193189 from 5 to 50 nM.
5. It is observed that different NPC lines might grow at slightly different rates. The optimal passaging ratio could range from 1:20 to 1:40. We recommend starting at 1:20 and then adjusting the seeding cell number to find the optimal passaging ratio in which NPCs can be stably passaged every 4 days.
6. NPCs are very sensitive to physical pipetting. When the NPCs are properly dissociated, less than five times pipetting up and down very gently is more than enough to dissociate the NPC aggregate into single cells. Too many pipetting will lead to cell death and loss of nephrogenic potential in the NPCs.
7. Medium in the U-bottom 96-well plates might evaporate fast. We avoid using the wells on the periphery of the 96-well plate, and instead, we fill these wells with 100 μL PBS in each well to prevent the culture medium from losing water too quickly. The loss of water in the culture medium leads to the change of osmosis and concentrations of NPSR components, which affects the NPC culture dramatically.
8. The gene-editing protocol described here is used as an example. Using different gene-editing tools other than CRISPR-Cas9, with different purposes of gene-editing, specific protocols should be developed accordingly. With the quick evolving of the gene-editing technology, better and more convenient protocols might be developed in the future following similar principles we describe here.

## Acknowledgments

This work was supported by Fundacion Dr. Pedro Guillen, Universidad Catolica San Antonio de Murcia, G. Harold and Leila Y. Mathers Charitable Foundation, The Leona M. and Harry B. Helmsley Charitable Trust, The Moxie Foundation, and National Institute of Health (R21 AG055938).

### References

1. Kobayashi A et al (2008) Six2 defines and regulates a multipotent self-renewing nephron progenitor population throughout mammalian kidney development. Cell Stem Cell 3:169–181
2. Diep CQ et al (2011) Identification of adult nephron progenitors capable of kidney regeneration in zebrafish. Nature 470:95–100
3. Nishinakamura R (2016) Advances in generating and maintaining nephron progenitors. Nat Rev Nephrol 12:67–68
4. Costantini F, Kopan R (2010) Patterning a complex organ: branching morphogenesis and nephron segmentation in kidney development. Dev Cell 18:698–712
5. Dressler GR (2006) The cellular basis of kidney development. Annu Rev Cell Dev Biol 22:509–529
6. Little MH, McMahon AP (2012) Mammalian kidney development: principles, progress, and projections. Cold Spring Harb Perspect Biol 4: a008300
7. Dudley AT, Godin RE, Robertson EJ (1999) Interaction between FGF and BMP signaling pathways regulates development of metanephric mesenchyme. Genes Dev 13:1601–1613
8. Barak H et al (2012) FGF9 and FGF20 maintain the stemness of nephron progenitors in mice and man. Dev Cell 22:1191–1207
9. Brown AC, Muthukrishnan SD, Oxburgh L (2015) A synthetic niche for nephron progenitor cells. Dev Cell 34:229–241
10. Tanigawa S, Taguchi A, Sharma N, Perantoni AO, Nishinakamura R (2016) Selective in vitro propagation of nephron progenitors derived from embryos and pluripotent stem cells. Cell Rep 15:801–813
11. Tanigawa S, Sharma N, Hall MD, Nishinakamura R, Perantoni AO (2015) Preferential propagation of competent SIX2+ nephronic progenitors by LIF/ROCKi treatment of the metanephric mesenchyme. Stem Cell Rep 5:435–447
12. Li Z et al (2016) 3D Culture supports long-term expansion of mouse and human nephrogenic progenitors. Cell Stem Cell 19:516–529
13. Costantini F, Watanabe T, Lu B, Chi X, Srinivas S (2011) Dissection of embryonic mouse kidney, culture in vitro, and imaging of the developing organ. Cold Spring Harb Protoc 2011: pdb.prot5613
14. Liao H-K et al (2015) Use of the CRISPR/Cas9 system as an intracellular defense against HIV-1 infection in human cells. Nat Commun 6:6413

# Part V

## 4D and 3D Ex Vivo Imaging of Embryonic Kidney Decellularization

# Chapter 14

# Time-Lapse Technologies and 4D Imaging of Kidney Development

Ulla Saarela and Ilya Skovorodkin

## Abstract

Time-lapse imaging is a technique of frequent imaging and following a course of a process. Because the development of the embryonic kidney can proceed ex vivo after dissection, it is possible to study the morphogenesis by culturing the kidney in the onstage incubator of a microscope and follow the developmental process by imaging. Confocal microscopes and other three-dimensional imaging systems offer the possibility for tracking the development process in four dimensions—3D and the time.

**Key words** Time-lapse imaging, Kidney development, Tissue culture

## 1 Introduction

Kidney development has been studied using time-lapse imaging already before digital cameras [1–3]. The digital cameras and the genetically modified mice have made the time-lapse techniques more available to study the morphogenesis in details. The GFP label on the cell membrane was introduced to study the branching of the ureteric bud [4]. Also patterning of the collecting duct system [5], behavior of cap mesenchymal cells [6], and self-organization after the dissociation of the kidney [7] have been studied using time-lapse imaging.

It is possible to apply cell segmentation and quantification in time-lapse images to detect individual cells and to follow their movement in the short-time range [8]. In this method, the embryonic kidney or kidney organoid is placed to grow in the restricted z-space making the explant grow in a flatten sheet of 70 μm thickness (Fig. 1). Because of the reduced thickness of the tissue, less z-planes are needed for the 3D imaging. In this system, there is only a cover glass between the imaged object and the objective, and this also improves the quality of the images and makes it possible to reduce the laser power, damaging the tissues less. The images can

Seppo Vainio (ed.), *Kidney Organogenesis: Methods and Protocols*, Methods in Molecular Biology, vol. 1926,
https://doi.org/10.1007/978-1-4939-9021-4_14, 

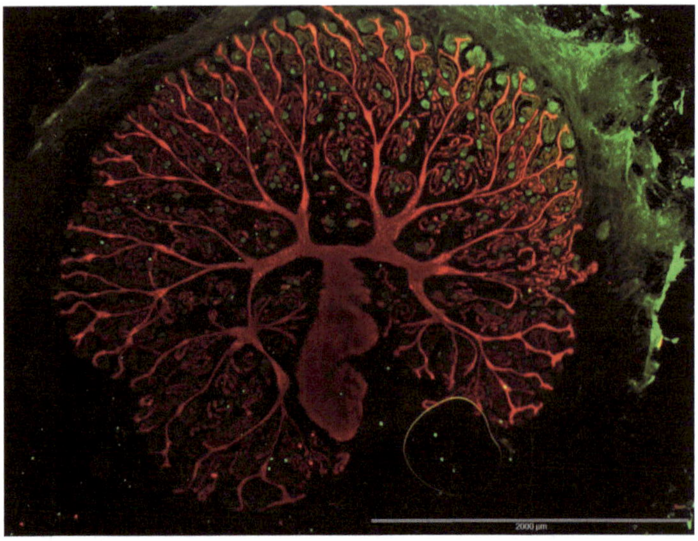

**Fig. 1** Embryonic mouse kidney cultured in the FiZD culture system for 12 days. Collecting ducts marked by Troma (red) and the glomeruli by Nephrin (green). Reproduced from Ref. [8] with permission from the Development

be taken with good resolution (voxel size $0.13 \times 0.13 \times 1.13$ μm) and with high frequency, at least every 5 min.

## 2 Materials

### 2.1 Fixed Z-direction Culture

1. 6-well Greiner CELLSTAR® multiwell culture plates (M9062 Sigma).
2. 24 mm Transwell® with 0.4 μm pore polyester membrane insert (CLS 3450, Sigma).
3. 70 μm polystyrene beads (100263-10, Corpuscular).
4. Matrigel® (Corning)—optional.
5. Histoacryl® glue (Braun Ref 1050052).
6. $24 \times 24$ mm cover glasses (cleaned according to the protocol).
7. 1 M HCl.
8. Dulbecco's Modified Eagle's Medium (DMEM, D777 Sigma).
9. Fetal bovine serum (FBS, Gibco).
10. Penicillin and streptomycin (P/S, Sigma).

## 3 Methods

### 3.1 Preparing the Plate for Imaging

1. Drill 20 mm in diameter holes in the bottom of wells (Fig. 2, *see* **Note 1**).
2. Smoothen the rims of the holes by rubbing with a pumice.

**Fig. 2** A modified 6-well plate with cover glass glued in the bottom of the wells. A Transwell® insert placed in well 2

3. Glue the cleaned coverslip on top of the hole using the Histoacryl® glue.
4. Before using rinse with ethanol, distilled water, and dry in a UV hood.

### 3.2 Cleaning of the Cover Glass

1. Heat coverslips in a loosely covered glass beaker in 1 M HCl at 50–60 °C for 4–16 h.
2. Cool to room temperature.
3. Rinse out 1 M HCl with $ddH_2O$.
4. Fill container with $ddH_2O$ and sonicate in water bath for 30 mins.
5. Repeat **step 4**.
6. Repeat **step 4**.
7. Fill container with 50% EtOH and 50% $ddH_2O$, and sonicate in water bath for 30 min.
8. Fill container with 70% EtOH and 30% $ddH_2O$, and sonicate in water bath for 30 min.
9. Fill container with 95% EtOH and 5% $ddH_2O$, and sonicate in water bath for 30 min.
10. Fill container with 95% EtOH.
11. Keep coverslips on Whatman filter paper in a box, and keep for autoclave.

### 3.3 Methods

#### 3.3.1 Fixed Z-Direction (FiZD) Culture System

1. Dissect kidneys and keep in the culture medium (*see* **Note 2**).
2. Mix an aliquot of polystyrene beads in Matrigel (on ice), and make 6–7 drops of Matrigel on the lower side of Transwell insert (chilled on ice) (*see* **Note 3**).

3. Add the kidneys on the drop of Matrigel or on the lower side of the Transwell insert. In the latter case, add the polystyrene beads at the bottom of the well on the cover glass (*see* **Note 4**).
4. Turn the insert around and transfer it into the well very gently.
5. Compress the kidney samples very slowly and gently until they are 70 μm in depth (control under the microscope).
6. When the samples are assembled, fix the inserts in the wells by melting the plastic at three positions with hot glass or metal needle.
7. Add about 2 mL of culture medium (DMEM/20% FBS/1% P/S) to the well. Place to the incubator at 37 °C and 5% $CO_2$.

*3.3.2 Time-Lapse Imaging*

1. Turn on the microscope (Zeiss LSM 780 confocal microscope) and the onstage incubator. Wait until the incubator temperature and carbon dioxide levels have settled to 37 °C and 5% $CO_2$.
2. Place the 6-well plate with the samples in the incubator.
3. Determine the position of the imaged samples and the imaging settings.
4. Determine the number and position of z-stacks.
5. Start the time-lapse imaging. Imaging set for every 5 to 20 min.
6. Save the files daily.

*3.3.3 Image Processing*

1. Deconvolute the data with Huygens Professional (Scientific Volume Imaging).
2. Combine the image data from different files to include the whole imaging time.
3. Further enhancement of the image quality and the segmentation of the data can be done using Matlab®(Mathworks)-based programs or Fiji [9].

## 4 Notes

1. Wear protective glasses and respiration filter because of small plastic shreds.
2. The culture method can be used also for kidney organoids and other tissues such as gonads [10].
3. Keep the plate also on ice to chill the glass bottom.
4. Matrigel is not necessary to use.

## Acknowledgments

This work was supported financially by the Suomen Akatemia (Academy of Finland) (206038, 121647, 250900, 260056; Centre of Excellence grant 2012-2017 251314), Munuaissäätiö—Finnish Kidney and Liver Association, the Sigrid Juséliuksen Säätiö, Novo Nordisk, Syöpäjärjestöt (Cancer Society of Finland), the European Community's Seventh Framework Programme (FP7/2007-2013; grant FP7-HEALTH-F5-2012-INNOVATION-1 EURenOmics 305608), and H2020 Marie Skłodowska-Curie Actions Innovative Training Network "RENALTRACT" Project ID 642937.

### References

1. Saxen L, Wartiovaara J (1966) Cell contact and cell adhesion during tissue organization. Int J Cancer 1(3):271–290
2. Saxén L, Toivonen S, Vainio T, Korhonen P (1965) Untersuchungen über die tubulogenese der niere. Zeitschrift Für Naturforschung B 20:4. https://doi.org/10.1515/znb-1965-0415
3. Saxén L (1987) Organogenesis of the kidney, 1st edn. Cambridge University Press, Cambridge
4. Costantini F, Watanabe T, Lu B, Chi X, Srinivas S (2011) Imaging kidney development. Cold Spring Harb Protoc 2011(5):pdb.top109. https://doi.org/10.1101/pdb.top109
5. Lindstrom NO, Chang CH, Valerius MT, Hohenstein P, Davies JA (2015) Node retraction during patterning of the urinary collecting duct system. J Anat 226(1):13–21. https://doi.org/10.1111/joa.12239
6. Combes AN, Lefevre JG, Wilson S, Hamilton NA, Little MH (2016) Cap mesenchyme cell swarming during kidney development is influenced by attraction, repulsion, and adhesion to the ureteric tip. Dev Biol 418(2):297–306. https://doi.org/10.1016/j.ydbio.2016.06.028
7. Lefevre JG, Chiu HS, Combes AN, Vanslambrouck JM, Ju A, Hamilton NA, Little MH (2017) Self-organisation after embryonic kidney dissociation is driven via selective adhesion of ureteric epithelial cells. Development 144 (6):1087–1096. https://doi.org/10.1242/dev.140228
8. Saarela U, Akram SU, Desgrange A, Rak-Raszewska A, Shan J, Cereghini S et al (2017) Novel fixed z-direction (FiZD) kidney primordia and an organoid culture system for time-lapse confocal imaging. Development 144(6):1113–1117. https://doi.org/10.1242/dev.142950
9. Schindelin J, Rueden CT, Hiner MC, Eliceiri KW (2015) The ImageJ ecosystem: an open platform for biomedical image analysis. Mol Reprod Dev 82(7–8):518–529. https://doi.org/10.1002/mrd.22489
10. Prunskaite-Hyyrylainen R, Skovorodkin I, Xu Q, Miinalainen I, Shan J, Vainio SJ (2016) Wnt4 coordinates directional cell migration and extension of the mullerian duct essential for ontogenesis of the female reproductive tract. Hum Mol Genet 25(6):1059–1073. https://doi.org/10.1093/hmg/ddv621

# Chapter 15

# Decellularizing and Recellularizing Adult Mouse Kidneys

Theresa Chow, Jennifer Whiteley, and Ian M. Rogers

## Abstract

Decellularization is the process by which resident cells are lysed and cellular debris is removed from the tissue, leaving behind the extracellular matrix scaffold. The extracellular matrix scaffold can be used for three-dimensional culturing of cells. Here we describe methods of decellularizing whole and thick sections of mouse kidneys using a 0.1% sodium dodecyl sulfate (SDS) detergent solution and strategies to repopulate whole and thick sections of decellularized mouse kidneys with cells.

**Key words** Mouse kidney dissection, Cannulation and perfusion, Kidney decellularization, Extracellular matrix, Kidney recellularization, Bioreactor, Three-dimensional (3D) tissue culture

## 1 Introduction

Organogenesis, the study of the development of organs, is best studied in vivo. Tissue transparency has allowed researchers to study amphibian and fish organ development in vivo. However, higher-order mammals do not have this tissue property, and thus the study of mammalian organogenesis must rely on alternative ex vivo methods. Dr. Lindbergh first described an organ culture system in 1935 which was then used by Dr. Carrel in 1937 for the in vitro culture of the thyroid gland [1, 2]. The culture of chicken or amphibian organ rudiments or dissociated cells was reported in the 1930s and 1940s, but organ cultures with mammalian cells did not occur until the 1950s [3]. The development of ex vivo explant culture has helped us better understand kidney organogenesis. However, explant cultures are limited by primary tissue availability and size since they rely on diffusion of medium and nutrient throughout the tissue.

The development of human embryonic stem cells [4] and induced pluripotent stem cells [5] and human organ cultures, referred to as organoid cultures, has helped advance the study of mammalian organ development. Organoids are made by culturing progenitor cells as a cell aggregate under conditions that mimic the

Seppo Vainio (ed.), *Kidney Organogenesis: Methods and Protocols*, Methods in Molecular Biology, vol. 1926,
https://doi.org/10.1007/978-1-4939-9021-4_15, 

in vivo environment. Recently, organoid culture has been used to generate kidney organoids from human pluripotent stem cells [6–8]. Interestingly, the organoids had a gene expression profile similar to first trimester fetal kidneys, indicating that the maturation of a human organoid runs in synchrony with normal human fetal development. Despite the power of organoid cultures, there are some limitations such as the lack of full organ organization and orientation. Furthermore, there is a size limitation to organoids as they are not vascularized in vitro [9].

Decellularization of organs to produce extracellular matrix scaffolds for stem cell differentiation is an alternative to organoids. Acellular tissue preparations were first described in 1948 [10], and early decellularization procedures in the 1960s were used to study the extracellular matrix content [11]. Extracellular matrix proteins are vital to normal organ development and tissue homeostasis. The extracellular matrix acts as a support for cells during embryo development, induces cell differentiation and proliferation, promotes cell adhesion and cell migration, and provides cues to maintain tissue homeostasis and regeneration. The extracellular matrix proteins are conserved between species, but the composition differs greatly between tissue and organs within an individual. Even within a single organ, the composition of the extracellular matrix is not uniform throughout and confers location-specific coordinates within the organ [12, 13]. Importantly, the extracellular matrix acts as a repository for growth factors. It is these properties of the acellular organ that make this substrate amenable to the generation of organs in a dish.

Decellularization to generate extracellular matrix substrates for tissue engineering started in the 1990s with the decellularization of rat aortic explants to generate substrates to study smooth muscle cell properties [14]. This was followed by the decellularization of simple tissues such as the trachea and bladder and then sections of cardiac tissue. Bladder reconstruction using decellularized matrices was reported in the mid-1990s [15]. Trachea decellularization and reconstruction was reported in the 2000s [16], and skin reconstruction using decellularized dermis was reported in 1991 [17]. Similar to organoids, acellular organ substrates combined with patient-derived stem cells can create a platform for personalized medicine studies. Properly decellularized organs retain both the structural proteins and biologically active growth factors that are capable of inducing location-specific differentiation of stem and progenitor cells [12, 13]. Stem cells and their differentiated products can orient and organize themselves using biological cues from the extracellular matrix proteins. Whole organ decellularization was pioneered by Doris Taylor's group in 2008 when they decellularized a whole rat heart [18]. In their groundbreaking study, they demonstrated that a solution composed of 1% SDS can successfully remove all the cells from a rat heart and leave behind an intact extracellular matrix that could support the growth of donor cells.

Although 3D-printed scaffolds are attractive alternatives to decellularized scaffolds, it is currently difficult to use this technology to replicate the intricate extracellular matrix structure of complex organs, such as the kidney, and the composition and levels of region-specific growth factors on the extracellular matrix scaffold. Therefore, 3D-printed scaffolds would provide a structural framework for the new organ but not the region-specific signals that direct the differentiation of stem and progenitor cells; these signals are required for producing highly organized organs such as the heart, lung, liver, pancreas, and kidney. Our group and others have demonstrated that adult decellularized scaffolds can support and direct the differentiation of stem and progenitor cells and neonatal cells to mature cell types [12, 13, 18–20]. Decellularized extracellular matrix scaffolds, with its built-in biological cues, can also guide and organize cells into nephrons that resemble the native kidney. Moreover, the intact vasculature in whole acellular kidneys allows for easy nutrient delivery into the kidney. Altogether, decellularized extracellular matrix scaffolds combined with stem and progenitor cells provide a unique platform and valuable tool to study kidney organogenesis in vitro.

## 2 Materials

### 2.1 Kidney Collection

1. Mice (any age).
2. Thumb forceps.
3. Blunt/blunt scissors.
4. Phosphate buffered saline—no $Ca^{2+}$ or $Mg^{2+}$ (PBS).

### 2.2 Tissue Dissection and Cannulation

1. 10 cm Petri dish.
2. Blunt/sharp scissors.
3. Spring scissors.
4. Thumb forceps.
5. Dumont #5 forceps.
6. 5-0 silk suture.
7. Sharpening stone.
8. 30 gauge × ½ in. needles.
9. Thin wall tubing.
10. General use silicone.

### 2.3 Kidney Decellularization

1. Peristaltic pump with two-stop tubing.
2. Small bore tubing.
3. Female luer lock.

4. Male luer lock.
5. Clamp and stand.
6. 1 cc syringe.
7. PBS.
8. 0.1% sodium dodecyl sulfate (SDS) solution: 0.1% w/v SDS dissolved in distilled water.
9. Penstrep.

### 2.4 Immunohistochemistry

1. 10% neutral buffered formalin (NBF).
2. Blocking buffer: PBS containing 0.1% Triton X-100 and 10% normal goat serum (NGS).
3. Wash buffer: PBS containing 0.1% Triton X-100.
4. Staining buffer: PBS containing 0.1% Triton X-100 and 1% normal goat serum (NGS).
5. Primary antibodies: laminin (Abcam, ab11575), fibronectin (Abcam, ab23750), collagen IV (Abcam, ab6586).
6. Secondary antibodies: anti-rabbit 488 (Invitrogen).
7. 4′,6-Diamidino-2-phenylindole (DAPI).
8. 1,4-Diazabicyclo [2.2.2] octane (DABCO) (Sigma).

### 2.5 Recellularization of Whole Acellular Kidney

1. 1 cc syringe.
2. Peristaltic pump.
3. 100 mL Corning glass jar with lid.
4. Small bore tubing.
5. General use silicone.
6. Three-way stopcocks.
7. Female luer lock.
8. Male luer lock.
9. 0.22 μm syringe filter.

### 2.6 Decellularization of Kidney Thick Sections

1. Scalpel blade.
2. Vibratome (optional).
3. Low melting agarose (optional).
4. 0.1% sodium dodecyl sulfate (SDS) solution: 0.1% w/v SDS dissolved in distilled water.

### 2.7 Recellularization of Kidney Thick Sections

1. 1 cc syringe.
2. 30 gauge × ½ in. needles.
3. Transwell insert.
4. 24-well plate.

## 3 Methods

### 3.1 Kidney Collection

Animals were cared for and handled in accordance with the Canadian Council on Animal Care and institutional guidelines (Toronto Centre for Phenogenomics).

1. Euthanize mice using $CO_2$.
2. Mice can be used at any age, but care must be taken to remove the visceral fat that accumulates in older or obese mice.
3. We do not perform cardiac perfusion prior to tissue collection as we found it was not necessary for successful decellularization. However, others have found that performing cardiac perfusion improves decellularization success rate. For training purposes, cardiac perfusion should not be done as the presence of blood in the vessels significantly improves visualization.
4. Pin down euthanized mice in a dorsal recumbent position. Using blunt/blunt scissors, make a midline incision from the pubis to the xiphoid process. Create a flap on each side by cutting from the midline over to the lateral side of the mouse at the top and bottom of the midline incision.
5. Using thumb forceps, dissect away the large and small intestines, stomach, and liver to expose the two kidneys.
6. Once kidneys are exposed, bisect the abdominal aorta and vena cava above and below the two kidneys. With a pair of blunt/blunt scissors, cut along the tissue below the artery and vein to excise both kidneys. Each kidney can be excised separately, but we find that it is better to remove the kidneys together and then dissect them apart under a microscope. This method will ensure that each kidney will have a sufficient length of blood vessel for cannulation.
7. Locate each ureter of the bladder and trace the ureter back to each kidney. Cut the ureter using blunt/blunt scissors to reduce tissue damage.
8. Place tissue in PBS on ice for transport to dissecting lab.

### 3.2 Cannula Production

1. Cut 10 cm lengths of thin wall tubing.
2. Slightly blunt 30 g needles using a sharpening stone.
3. Add a small amount of general use silicone to the length of the needle avoiding the eye of the needle.
4. Thread the needle into the thin wall tubing; dap a small amount of the silicone around the hub of the needle to secure the tubing (Fig. 1).
5. Hang cannula upright to allow silicone to dry.
6. Autoclave to sterilize cannula.

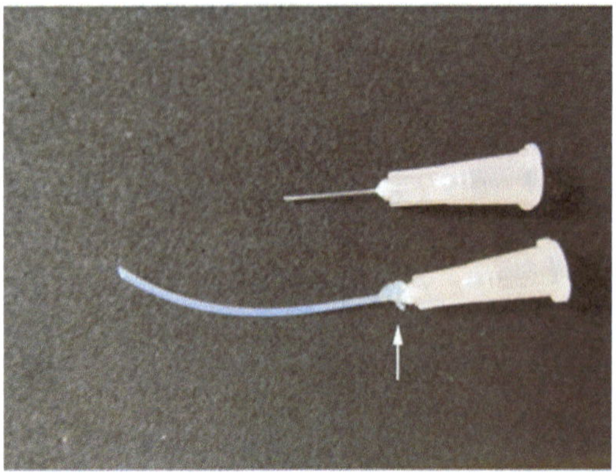

**Fig. 1** Cannula production. A blunted 30 g needle is inserted into a piece of thin wall tubing to produce a soft cannula. Silicone (arrow) is applied to hold the tubing over the blunted needle

### *3.3 Tissue Dissection and Cannulation for Whole Organ Decellularization*

1. Place kidneys in a 10 cm petri dish with PBS and on dissecting microscope.
2. Using Dumont forceps, blunt dissect visceral fat from the surrounding tissue to improve visualization of the blood vessels (Fig. 2).
3. Using blunt/sharp scissors, separate the kidneys as close to the descending aorta as possible to allow for a sufficient length of blood vessel to cannulate. The descending aorta can be left on one of the kidneys as this landmark helps with cannulation.
4. Blunt dissect surrounding tissue from the artery.
5. Place a piece of 5-0 suture around the artery. It should be placed far enough down the vessel to accommodate the eye of the cannula but still outside of the kidney. Pre-tie a double knot to make tightening the suture easier.
6. Using spring scissors, make a clean cut on the artery. If the artery frays with multiple attempts to cannulate, recut the vessel. It is much easier to cannulate with a clean cut.
7. Cannulate the renal artery using a lab prepared cannula consisting of a blunted 30 g needle and thin wall tubing. Alternatively, a blunted 30 g needle can be used to directly cannulate the artery, but the flow rate has to be reduced to prevent the kidney from falling off the needle. Cannulas and needles should be preloaded with PBS using a 1 cc syringe to reduce air pockets forming in the kidney and damaging the tissue.
8. Grasp the tissue surrounding the vessel, or grasp one edge of the vessel with Dumont forceps, and advance the cannula into the artery (*see* **Note 1**).
9. Once the cannula is inserted, gently set the cannula down to tie the suture. Grasp the two ends of the suture with the double

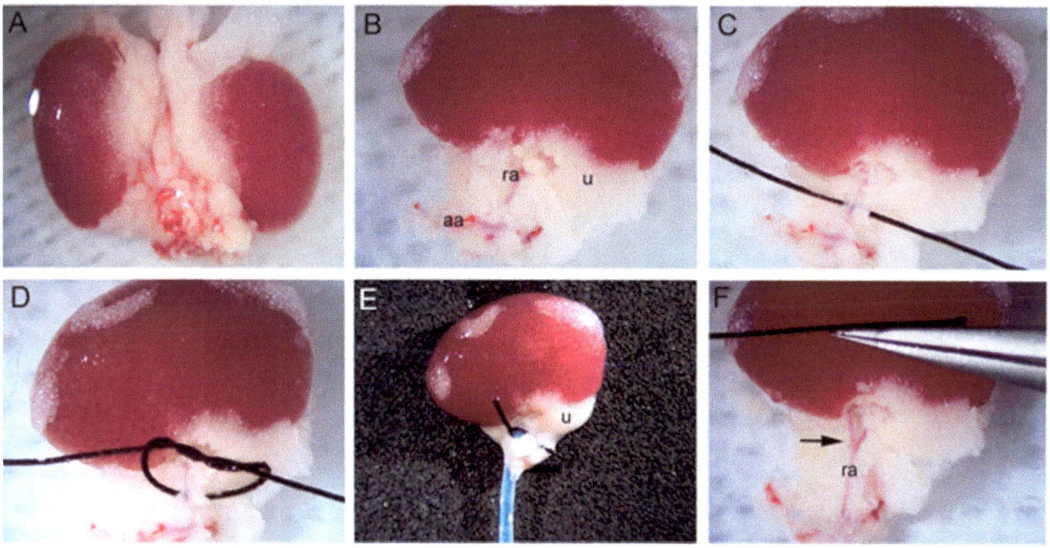


**Fig. 2** Kidney dissection and cannulation. (**a**) Kidneys prior to any blunt dissection. (**b**) Kidneys are separated and fat is removed to expose the vessels. (**c**) A piece of suture is placed under the renal artery. (**d**) A double knot is loosely tied around the renal artery. (**e**) Soft cannula is inserted into the renal artery, and suture is tied and locked with another single knot. Ensure that the renal vein is not tied off as the kidney will not drain during decellularization. More blunt dissection is done past the suture knot to remove as much tissue as possible. (**f**) Renal artery is stretched to show the bifurcation. Inserting the cannula beyond the bifurcation will result in partial decellularization. *ra* renal artery, *aa* abdominal aorta, *u* ureter

knot already loosely tied, and tighten the knot making sure the eye of the cannula is advanced past the knot. Lock the double knot with a single knot.

10. To cannulate the ureter, visualize the ureter as it enters the kidney, and blunt dissect the tissue away to allow for placement of the suture. Pre-tie a double knot in the suture. The ureter is fragile and its diameter increases as it gets close to the kidney so it's easier to cannulate closer to the kidney.
11. Make a clean cut on the ureter using spring scissors, insert the cannula, and tighten the suture. The ureter can be cannulated before or after decellularization.

### 3.4 Whole Kidney Decellularization

1. Place tubing on the pump head according to the pump operating instructions. Be sure to calibrate the pump to the size of tubing being used. We use two-stop peristaltic pump tubing and attach additional lengths of tubing using male and female luer locks to the pump tubing to reach into our prepared solutions. It is important to make sure that all the tubing throughout the pump system has the same inner diameter to ensure that the flow rate is constant throughout the system.
2. Hang the tubing over a generic lab stand with a clamp, and place a beaker under the tubing to collect the discard (Fig. 3).

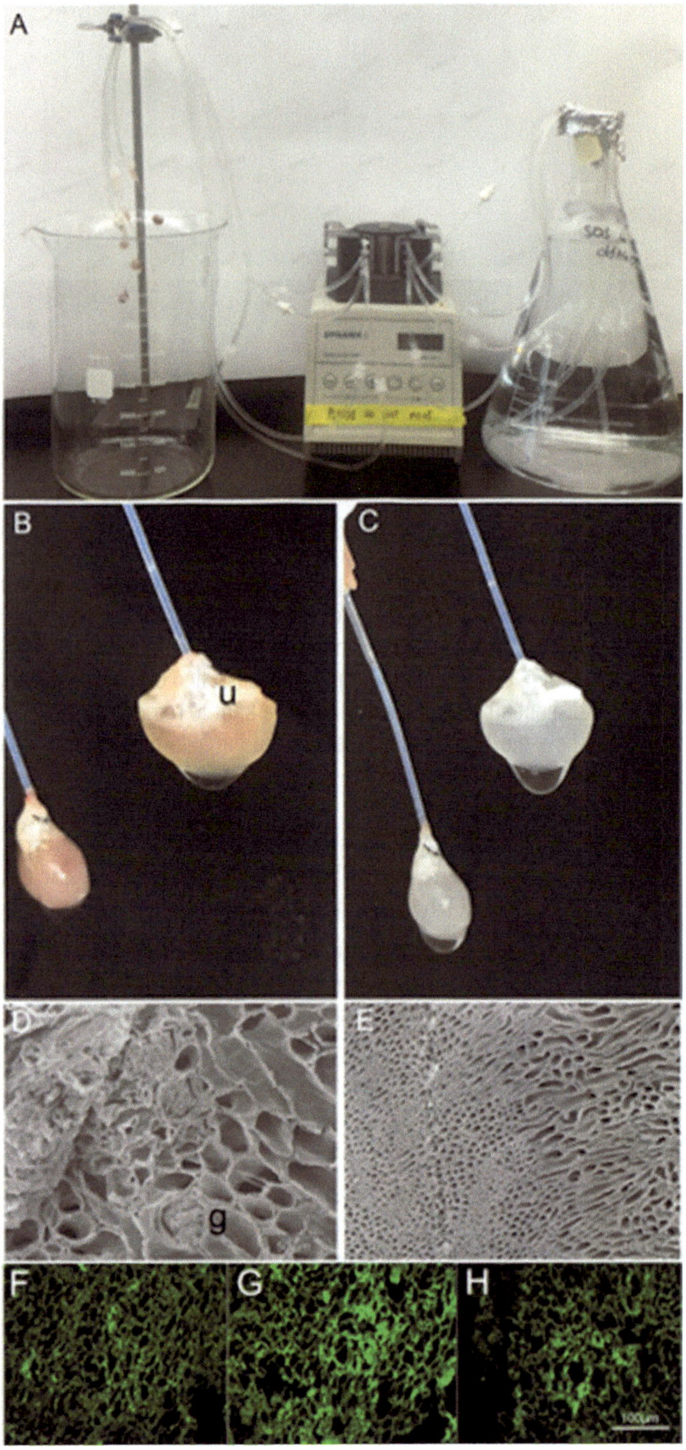


**Fig. 3** Decellularization setup. (**a**) A peristaltic pump is used to deliver the detergent solution into the renal artery. (**b**) Kidneys after 4 h of decellularization (0.4 mL/min). (**c**) Kidneys after 48 h of decellularization (0.4 mL/min). (**d**, **e**) Scanning electron microscopy (SEM) of the acellular kidney. *g* glomerulus. (**f–h**) Immunohistochemistry of extracellular matrix proteins, (**f**) fibronectin; (**g**) laminin; (**h**) collagen IV

3. Place a male luer lock at the end of the tubing to attach the needle/cannula.
4. Preload the tubing with 0.1% SDS detergent solution.
5. Once all tubing is primed and loaded, with all bubbles removed, attach the cannulated kidney to the male luer lock.
6. Run the 0.1% SDS solution through the kidneys at 0.4 mL/min for 48–72 h (*see* **Note 2**). Smaller kidneys will clear within 48 h, while larger kidneys can take up to 72 h (*see* **Note 3**). If a blunted 30 g needle is used for cannulation instead of a cannula, then reduce the flow rate to 0.2 mL/min to prevent the kidney from slipping off the needle (*see* **Note 4**).
7. Following SDS treatment, wash kidneys with PBS containing 1× Penstrep for 12–24 h to reduce possible contamination. Using sterile equipment and cannula and performing the decellularization in a fume hood help to reduce contamination during the decellularization process. There are sterilization techniques available, but they may compromise the integrity of the extracellular matrix proteins (*see* **Note 5**).

### 3.5 Characterization of Acellular Kidney

Decellularization must be worked out empirically. Flow rate, decellularization time, tubing size, and kidney size can all impact efficiency. Each parameter must be tested to determine the optimal decellularization condition. To assess for decellularization success, perform histological staining to confirm the absence of cells and debris and retention of extracellular matrix proteins. DAPI and hematoxylin and eosin (H&E) staining can be performed to assess for the presence of cells and nuclear debris. Scanning electron microscopy (SEM) can be used to examine the ultrastructure of the decellularized kidney. Immunostaining for collagen, laminin, and fibronectin along with Alcian Blue staining can be used to detect the presence of basement membrane proteins and glycosaminoglycan (GAG) chains.

#### 3.5.1 Immunohistochemistry for Fibronectin, Laminin, and Collagen IV (See Notes 6 and 7)

1. Fix acellular kidneys in 10% NBF for 2–3 h at room temperature.
2. Dehydrate tissues following standard protocols.
3. Embed fixed and dehydrated tissues in paraffin following standard protocols.
4. Section 5-μm-thick tissue sections using microtome.
5. Dewax and rehydrate tissue sections following standard protocols.
6. Perform antigen retrieval (if required).
7. Incubate tissue sections in blocking buffer for 1 h at room temperature.

8. Wash slides once in wash buffer.
9. Dilute primary antibody in staining buffer and apply onto tissue sections. Incubate overnight at 4 °C.
10. Wash slides four times in wash buffer (10 min each time).
11. Dilute secondary antibody in staining buffer and apply onto tissue sections. Incubate for 1 h at room temperature in the dark.
12. Wash slides four times in wash buffer (10 min each time).
13. Stain tissue sections with DAPI.
14. Wash slides once in wash buffer.
15. Coverslip slides with DABCO and seal.

#### 3.5.2 Nucleic Acid Detection

1. Stain acellular kidney sections with DAPI to detect residual nuclei acids (*see* **Note 8**) [21].
2. DNA Quantitation Assay: Mince tissue and place in a mortar with liquid nitrogen. Using a pestle, grind the tissue to a powder. After liquid nitrogen is evaporated, DNA is extracted using a chloroform/phenol extraction method and measured by UV absorbance [12].

#### 3.5.3 Hematoxylin and Eosin (H&E) and Alcian Blue Stains

1. Stain acellular kidney sections with H&E following standard protocols to detect residual nuclei and cells [22].
2. Stain acellular kidney sections with Alcian Blue following standard methods to detect glycosaminoglycan (GAG) chains [23].

#### 3.5.4 Scanning Electron Microscopy (SEM)

Prepare acellular tissue for SEM following standard protocols. SEM will allow for the assessment of physical damage to the extracellular matrix and for residual cells [13].

### 3.6 Recellularization of Whole Acellular Kidneys: Simple Syringe Method

1. Attach a 1 cc syringe to the needle/cannula used to cannulate the renal artery.
2. Repopulate the nephrons and collecting duct by resuspending experimental cells in approximately 0.5 mL of medium, and load cell suspension into a 1 cc syringe (*see* **Note 9**). Attach syringe to the cannula used to cannulate the ureter.
3. Slowly draw the plunger of the syringe attached to the artery while pushing down on the plunger of the syringe attached to the ureter. This must be done carefully to avoid damaging the kidney (*see* **Note 10**).
4. Repopulate the vasculature by resuspending the experimental cells in 0.5 mL medium, loading the cells into a 1 cc syringe, and gently injecting them into the renal artery.
5. Replace medium in the repopulated kidney with fresh medium (1 mL) every day through the renal artery.

### *3.7 Recellularization of Whole Acellular Kidneys: Vacuum Chamber Method*

#### *3.7.1 General Setup of Vacuum Chamber*

1. Drill four holes into the lid of a 100 mL Corning glass jar (Fig. 4). The holes will be used to create four ports:
   (a) Port #1: Ureter connection for loading cells or measuring urine during organ growth.
   (b) Port #2: Vascular connection (artery) for loading cells or medium circulation (IN) during organ growth.
   (c) Port #3: Vacuum gauge port during cell loading or vascular/medium circulation (OUT) during organ growth.
   (d) Port #4: Vacuum line in port during cell loading or air supply port during organ growth.
2. Cut 2 cm lengths of small bore tubing, and push a piece of tubing through each of the four holes. Use tubing that is slightly larger than the holes, and/or apply general use silicone to seal any gap between the hole and tubing.
3. Attach male luer locks to the end of the tubing facing inside the glass jar and female luer locks to the other end.
4. Attach three-way stopcocks to the female luer locks for better control of medium/air flow in and out of the chamber. The stopcocks can also be used to block female luer locks when they are not in use.

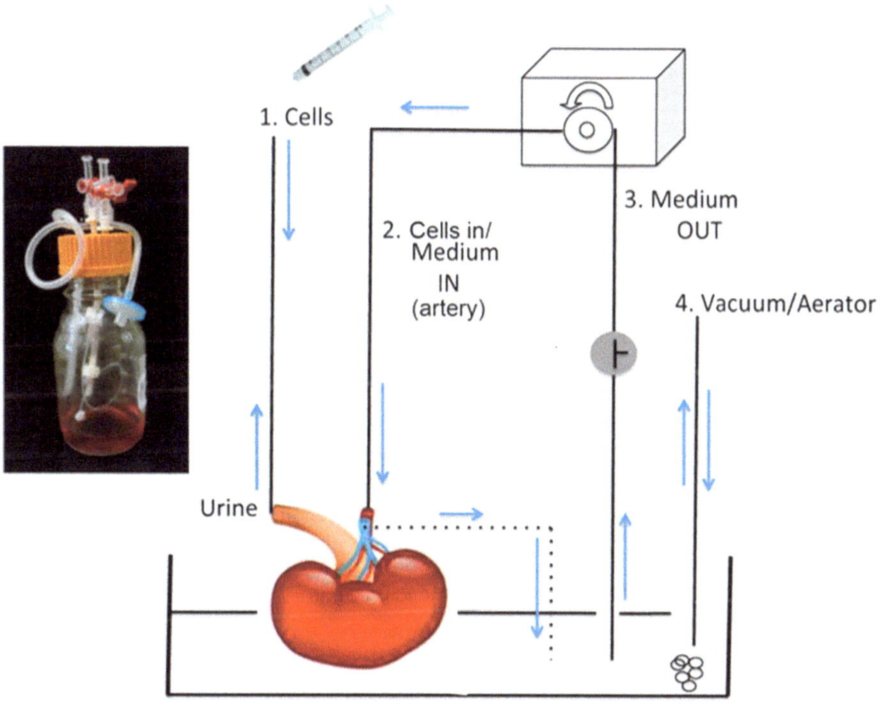


**Fig. 4** Vacuum and growth chamber setup. The four ports are used for different purposes, depending on whether the chamber is being used for cell loading or for organ growth. The arrows show the flow of medium, cells, air, or vacuum depending on the use of the chamber

5. Attach the ureter cannula to the male luer lock of port #1.
6. Attach the renal artery cannula to the male luer lock of port #2.
7. Attach a pressure gauge to port #3.
8. Attach a vacuum line to port #4.

*3.7.2 Loading Cells into the Ureter*

1. Resuspend experimental cells in approximately 0.5 mL medium, and load cell suspension into a 1 cc syringe (*see* **Note 9**).
2. To repopulate the nephrons and collecting duct, attach the syringe with cells to the stopcock of port #1.
3. Close off port #2 using the stopcock.
4. Apply vacuum (−30 mmHg maximum) to the airtight chamber. As the vacuum is applied, the syringe plunger will depress on its own as the cells are drawn into the kidney. Hold the vacuum pressure constant for 2 min, and then slowly release the vacuum (*see* **Notes 10** and **11**).
5. Remove the syringe from the stopcock and close off port.

*3.7.3 Loading Cells into the Vasculature*

1. Resuspend experimental cells in approximately 0.5 mL medium, and load cell suspension into a 1 cc syringe.
2. To repopulate the vasculature, attach the renal artery cannula to the male luer lock of port #2, and attach the syringe with cells to the stopcock above it.
3. Slowly and gently press down on the plunger to inject cells into the renal artery.
4. Remove syringe from the stopcock.

*3.7.4 Setting Up Vacuum Chamber for Organ Culture*

1. Remove the vacuum line from port #4, attach a 0.22 μm filter to the stopcock, and connect the port to an air supply or left open for passive air exchange. The 0.22 μm filter is added to filter the incoming air.
2. Remove the pressure gauge from port #3.
3. Attach one end of small bore tubing to the stopcock of port #2, pass tubing through a peristaltic pump, and connect the other end to the stopcock of port #3.
4. Attach another piece of small bore tubing to the male luer lock of port #3. This piece of small bore tubing should be long enough to touch the bottom of the Corning glass jar.
5. System is placed in an incubator (37 °C, 5% $CO_2$).
6. Add 50 mL of medium to the jar, and turn on peristaltic pump to circulate medium in to the kidney (*see* **Note 12**).
7. Medium can be sampled from port #3, and "urine" can be sampled from port #1.

### 3.8 Characterization of Recellularized Whole Kidneys

Tissue culture duration and conditions have to be optimized for each cell-scaffold system. Different cells used for recellularization will require different medium and culture conditions in order to adhere, proliferate, and mature on the scaffold. Similarly, kidneys decellularized with a different detergent will have different properties. Various histological staining and assays can be used to evaluate the recellularized kidney. DAPI and H&E staining can be used to help enumerate cell numbers and estimate cell dispersion within the scaffold and percent coverage. Immunohistochemistry staining using a combination of stem, progenitor, and mature markers in combination with scanning and transmission electron microscopy can be used to identify the cell types on the scaffold.

In addition to phenotypic characterization of the cells on the scaffold, it is also important to evaluate the function of the cells. Various kidney in vitro assays are available. Dextran or cisplatin can be added to the culture to assess the endocytic properties of proximal tubule cells. Song and colleagues [20] have described an in vitro method to assess if recellularized kidneys can clear creatinine, retain protein, reabsorb glucose and electrolytes, and produce concentrated urine. However, long-term transplantation and animal survival study are the only definitive functional test and will most likely require moving to large animal models.

### 3.9 Decellularization of Kidney Thick Sections

1. Follow **steps 1–8** in Subheading 3.1 Kidney Collection for obtaining a kidney.
2. Section the kidney into thick sections using scalpel or a vibratome. If using a vibratome, the kidney has to be embedded in 6% low melting agarose gel and sectioned at a speed of 0.3 mm/s.
3. Place thick sections in a large volume of 0.1% SDS detergent solution, and gently agitate the sections at room temperature.
4. Replace the 0.1% SDS detergent solution every 3–4 h in the first 48 h and every 6 h in the last 24 h of decellularization. Exact time for decellularization depends on tissue section thickness and should be determined empirically.
5. Alternatively, decellularized thick sections can be obtained by sectioning a whole decellularized kidney using a scalpel or vibratome.
6. Analyze decellularized thick sections using the same approach described for decellularized whole kidneys.

### 3.10 Recellularization of Kidney Thick Sections

1. Grow and prepare experimental cells for recellularization. Experimental cells can be primary adult or fetal kidney cells, renal cell lines, and/or stem cell-derived kidney cells. The number of cell to use for repopulation is to be determined empirically.

2. Resuspend experimental cells in approximately 0.1–0.2 mL medium, and load cell suspension into a 1 cc syringe. Attach a 30-gauge needle to the syringe.
3. Inject the cells into multiple areas of the decellularized thick section. Alternatively, cells can be layered on top of the thick section.
4. Recellularized thick sections can be cultured on air-liquid interface using Transwell inserts or submerged in medium in a tissue culture plate.
5. Analyze recellularized thick sections using the same approach described for recellularized whole kidneys.

## 4 Notes

1. The physiology of the kidney vasculature is sometimes different between individuals. Avoid advancing the cannula into the kidney. When the artery enters the kidney, the vessels bifurcate (*see* Fig. 2f). If the cannula is advanced past this bifurcation, then only one half of the kidney will decellularize.
2. The kidney is a dense tissue and not all detergent solutions will work. We experimented with different solutions but found that 0.1% SDS made with distilled water was optimal. Strong ionic detergents, such as SDS, are capable of either binding to extracellular matrix proteins even after extensive washing or altering protein configuration. Milder nonionic or zwitterionic detergents are less harsh than ionic detergents, and therefore extracellular matrix integrity is better preserved with these detergents, but they are ineffective at decellularizing dense tissues such as the kidney. We have found that the low 0.1% SDS solution does not hinder the growth or differentiation of nephron progenitor cells [13]. It has been reported that the use of a higher 1% SDS solution to decellularized hearts maintains extracellular matrix proteins, including glycosaminoglycan (GAG)-associated growth factors [18]. Optional: After SDS treatment, the kidney can be washed with 10× volume PBS followed by 1% Triton X-100 for 30 mins to help remove residual SDS [24].
3. Extended exposure to the detergent solution can result in loss of some extracellular matrix proteins. Whether or not the loss of proteins is critical is dependent on the experiment.
4. Depending on the size of the kidney and the experience of the person performing the cannulation, it may be easier to cannulate the renal artery using a blunted 30 g needle instead of a

cannula. If this is the case, then to prevent the kidney from sliding off the cannula during perfusion, the flow rate should be kept at 0.2 mL/min. At the slower flow rate, the decellularization normally takes 72 h. The soft cannula is more difficult to set up but allows for faster flow rates (up to 0.4 mL/min) and shorter decellularization times (48 h).

5. Alternatives to antibiotics for decontamination include 70% ethanol solution, 0.2% peracetic acid in 1 M NaCl, or radiation [25]. As each of these methods may alter the extracellular matrix proteins, we have chosen not to use them. We have found with mouse kidneys that increased flow rates during decellularization can eliminate cellular and nuclear debris, which reduces the risk of contamination.
6. As a control, native kidneys can be used to ensure reproducibility of the location of specific proteins, but this does not allow for quantitation of the remaining extracellular matrix proteins.
7. In our studies, although laminin and fibronectin and collagen IV were widely positive within the ECM, they showed different patterns of staining with some areas showing stronger staining than others.
8. Whole cells and residual nuclear debris can be detected using DAPI staining. Nuclear debris can be removed by increasing the flow rate during decellularization. Alternatively, benzonase endonuclease treatment can be used but it is expensive.
9. Experimental cells can be primary adult or fetal kidney cells, renal cell lines, and/or stem cell-derived kidney cells. The number of cell to use for repopulation is to be determined empirically.
10. There is little evidence that repopulated cells migrate through the acellular extracellular matrix scaffold in the short term. Therefore, cells used for repopulation of the nephron and collecting duct should be delivered through the ureter, while cells used for repopulation of the renal vasculature should be delivered through the renal artery. Simple injection of cells into ureter results in the accumulation of cells in the renal pelvis. Applying a vacuum to generate a pressure difference in the kidney will help distribute the cells more evenly throughout the kidney.
11. Do not damage the kidney capsule during cell loading. Damage to the kidney capsule will result in the collapse of the kidney.
12. Cell culture medium is dependent on the cell type being studied.

## References


1. Lindbergh CA (1935) An apparatus for the culture of whole organs. J Exp Med 62(3):409–431
2. Carrel A (1937) The culture of whole organs: I. Technique of the culture of the thyroid gland. J Exp Med 65(4):515–526
3. Moscona A, Moscona H (1952) The dissociation and aggregation of cells from organ rudiments of the early chick embryo. J Anat 86 (3):287–301
4. Thomson JA et al (1998) Embryonic stem cell lines derived from human blastocysts. Science 282(5391):1145–1147
5. Takahashi K, Yamanaka S (2006) Induction of pluripotent stem cells from mouse embryonic and adult fibroblast cultures by defined factors. Cell 126(4):663–676
6. Taguchi A et al (2014) Redefining the in vivo origin of metanephric nephron progenitors enables generation of complex kidney structures from pluripotent stem cells. Cell Stem Cell 14(1):53–67
7. Morizane R et al (2015) Nephron organoids derived from human pluripotent stem cells model kidney development and injury. Nat Biotechnol 33(11):1193–1200
8. Takasato M, Little MH (2017) Making a kidney organoid using the directed differentiation of human pluripotent stem cells. Methods Mol Biol 1597:195–206
9. Takasato M et al (2015) Kidney organoids from human iPS cells contain multiple lineages and model human nephrogenesis. Nature 526 (7574):564–568
10. Poel WE (1948) Preparation of acellular homogenates from muscle samples. Science 108(2806):390–391
11. Lidsky MD, Sharp JT, Rudee ML (1967) Studies on acellular bovine glomeruli. Isolation, chemical composition, and demonstration of collagen with an unusual hydroxylysine: hydroxyproline ratio. Arch Biochem Biophys 121(2):491–501
12. Shojaie S et al (2015) Acellular lung scaffolds direct differentiation of endoderm to functional airway epithelial cells: requirement of matrix-bound HS proteoglycans. Stem Cell Rep 4(3):419–430
13. Sambi M et al (2017) Acellular mouse kidney ECM can be used as a three-dimensional substrate to test the differentiation potential of embryonic stem cell derived renal progenitors. Stem Cell Rev 13(4):513–531
14. Battle T, Arnal JF, Michel JB (1994) Hyperproliferation of aortic smooth muscle cells and fibroblasts from young SHR rats is not shared by endothelial cells. Clin Exp Pharmacol Physiol 21(12):981–989
15. Sutherland RS, Baskin LS, Hayward SW, Cunha GR (1996) Regeneration of bladder urothelium, smooth muscle, blood vessels and nerves into an acellular tissue matrix. J Urol 156(2 Pt 2):571–577
16. Hung SH, Su CH, Lin SE, Tseng H (2016) Preliminary experiences in trachea scaffold tissue engineering with segmental organ decellularization. Laryngoscope 126(11):2520–2527
17. Krejci NC, Cuono CB, Langdon RC, McGuire J (1991) In vitro reconstitution of skin: fibroblasts facilitate keratinocyte growth and differentiation on acellular reticular dermis. J Invest Dermatol 97(5):843–848
18. Ott HC et al (2008) Perfusion-decellularized matrix: using nature's platform to engineer a bioartificial heart. Nat Med 14(2):213–221
19. Ross EA et al (2009) Embryonic stem cells proliferate and differentiate when seeded into kidney scaffolds. J Am Soc Nephrol 20 (11):2338–2347
20. Song JJ et al (2013) Regeneration and experimental orthotopic transplantation of a bioengineered kidney. Nat Med 19 (5):646–651
21. Kapuscinski J (1995) DAPI: a DNA-specific fluorescent probe. Biotech Histochem 70 (5):220–233
22. Kmiec Z (2016) J.A. Kiernan. Histological and histochemical methods: theory and practice, vol 54(1), 5th edn. Scion Publishing, Banbury, pp 58–59 571 pp. Folia Histochem Cytobiol
23. Schumacher U, Adam E (1994) Standardization of staining in glycosaminoglycan histochemistry: alcian blue, its analogues, and diamine methods. Biotech Histochem 69(1):18–24
24. Ossipow V, Laemmli UK, Schibler U (1993) A simple method to renature DNA-binding proteins separated by SDS-polyacrylamide gel electrophoresis. Nucleic Acids Res 21 (25):6040–6041
25. Poornejad N et al (2016) Comparison of four decontamination treatments on porcine renal decellularized extracellular matrix structure, composition, and support of human renal cortical tubular epithelium cells. J Biomater Appl 30(8):1154–1167

# Chapter 16

# Optical Projection Tomography Imaging to Study Kidney Organogenesis

Renata Prunskaite-Hyyryläinen

## Abstract

Optical projection tomography (OPT) is a 3D imaging technology. The 3D tomographic reconstruction permits precise analysis and quantification of various structures in developing embryonic tissues and adult organs of small rodents or biopsies. OPT enables detailed and accurate studies of kidney organogenesis, namely, ureteric tree branching morphogenesis and nephron quantification.

**Key words** Optical projection tomography, Phenotyping, 3D imaging, Branching morphology, Nephron quantification, Whole-mount immunostaining, Whole-mount in situ hybridization

## 1 Introduction

Optical projection tomography (OPT) is a three-dimensional (3D) imaging technology. It allows rapid 3D imaging of mesoscopic (1–10 mm) specimens that have fluorescent and nonfluorescent signals [1]. This technique has made a significant impact in the field of developmental biology. OPT permits acquisition of 3D digital data, which is used for computational analysis of multiple parameters such as developing kidney ureteric tree branch length and angles or nephron distribution and counts.

Prior the OPT scan, the specimen is stained and embedded to an agarose gel which serves as a support during imaging. The clearing agents are used to optically clear the specimen. This will enable light passage through the tissue with minimal scattering, absorption, and reflection. Cleared specimen is attached to a magnetic holder and steadily rotated 360 degrees around a single axis. A digital image is taken at every chosen angular step (0.45–0.9 degrees). After image acquisition, virtual sections are reconstructed by using a back-projection algorithm. The high-resolution images can be viewed and analyzed as tomographic sections and as a whole.

Seppo Vainio (ed.), *Kidney Organogenesis: Methods and Protocols*, Methods in Molecular Biology, vol. 1926,
https://doi.org/10.1007/978-1-4939-9021-4_16, © Springer Science+Business Media, LLC, part of Springer Nature 2019

Here will be described the (we describe) detailed step-by-step protocols for mouse embryonic kidney preparation for OPT after whole-mount immunostaining, whole-mount in situ hybridization (WISH), or both techniques combined.

Whole-mount immunostaining permits antibody staining and protein expression detection in the entire embryo or tissue. The WISH is using labeled complimentary RNA to localize specific RNA sequence in the tissue. There are numerous WISH protocols; here -in is presented (we will describe) an automated whole-mount in situ hybridization procedure [2]. Automated process allows completing the WISH within 3 days with moderate hands on time versus laborious 5-day manual protocol. The automated WISH protocol can be done manually using the same material and procedure. It will also be described how to combine WISH and immunostaining techniques allowing simultaneous examination of protein and RNA expression in the same tissue.

## 2 Materials

### 2.1 Whole-Mount Immunostaining

1. Phosphate buffered saline (PBS).
2. PBT-X-100: 1× phosphate buffered saline (PBS) with 0.1% Triton X-100 added. Make it fresh.
3. Paraformaldehyde (PFA) 4%: dissolve PFA powder in 1× PBS in water bath at 65 °C. Filter through 0.45 μm syringe filter.
4. Bleaching buffer: 15% $H_2O_2$ and 15% dimethyl sulfoxide (DMSO) dilute in absolute methanol, and filter through 0.45 μm syringe filter.
5. Blocking solution: 5% DMSO, 10% of heat-inactivated serum, derived from the same species as secondary antibody, diluted in PBT-X-100. Filter through 0.45 μm syringe filter.

#### 2.1.1 Whole-Mount In Situ Hybridization Probe Synthesis

1. RNAse-free diethylpyrocarbonate (DEPC)-treated water: 1/100 vol 10% DEPC in 100% EtOH; add double-distilled (dd) water and autoclave.
2. 5× transcription optimized buffer (Promega) comes with the RNA polymerase and contains DTT 100 mM (Promega).
3. Dig RNA labeling mix 10× Conc (Roche).
4. RNAse inhibitor, Recombinant RNasin® Ribonuclease Inhibitor 2500u, Promega.
5. Appropriate RNA polymerase: SP6, T3, or T7.
6. DNAse I (Promega).
7. 4 M LiCl: dissolve in DEPC water and autoclave.
8. 70% ethanol diluted in DEPC water.

*2.1.2 Whole-Mount In Situ Hybridization*

1. Diethylpyrocarbonate (DEPC) treated water made as in Subheading 2.1.1, **item 1**.
2. Phosphate buffered saline (PBS).
3. PBT-20: 1× PBS with 0.1% Tween 20. Make it fresh.
4. Hybridization buffer: 50% ultrapure formamide, 5× SSC (20× SSC stock solution: 175 g NaCl, 88.2 g Na citrate; set pH to 4.5 and add DEPC water up to 1 L), 50 μg/mL of tRNA, 1% SDS, 50 μg/mL heparin sodium salt Grade I-A, and DEPC water. Buffer can be prepared in advance and stored at −20 °C for a month.
5. Proteinase K stock solution: 10 mg/mL dissolved in sterile DEPC water; store at −20 °C. Working solution 0.01 mg/mL diluted in PBT-20.
6. Paraformaldehyde (PFA) 4%: dissolve PFA powder in 1× PBS prepared in DEPC-treated water, in water bath at 65 °C.
7. Glycine 0.2%: dissolve in PBT-20, make it fresh before use, and filter through 0.45 μm syringe filter.
8. Hydrogen peroxide 6%: dilute in DEPC water.
9. Washing buffer I: 25 mL formamide ≥99%; 12.5 20× SSC, pH 4.5 (made as in Subheading 2.1.2, **item 4** but use dd water instead of DEPC), 5 mL 10% SDS, 7.5 mL dd water.
10. Washing buffer II: 25 mL formamide ≥99%; 5 mL 20× SSC (made as in Subheading 2.1.2, **item 4** but use dd water instead of DEPC), 1 mL 10% SDS, 19 mL DEPC water.
11. Maleic acid buffer (MAB): 100 mM maleic acid, 150 mM NaCl, pH 7.5.
12. MABT: maleic acid buffer (MAB) with 0.1% of Tween 20. Add Tween 20 just before use.
13. Blocking solution: 0.5% blocking reagent (Roche) dissolved in MAB according to manufacturer's recommendations, supplemented with 10% heat-inactivated sheep serum.
14. Antibody and antibody dilution buffer: 0.5% blocking reagent (same as in **item 13**, Roche) dissolved in MAB, supplemented with 1% heat-inactivated sheep serum, and add anti-digoxigenin-AP Fab (Roche) diluted 1:1000 (1.5 U/mL).
15. NTMT: 100 mM NaCl, 100 mM Tris, pH 9.5, 50 mM $MgCl_2$, 1% Tween 20.
16. BM Purple AP Substrate precipitating solution (Roche).
17. InsituPro device from Intavis AG Bioanalytical Instruments or equivalent.

***2.2 Preparation for OPT Scanning***

1. A 5 mL syringe.
2. Regular or low melting point agar.

3. Clearance solution BABB: benzyl alcohol and benzyl benzoate solution dilution ratio 1:2.
4. Clearance solution scaleCUBIC-1 reagent [3]: 25 wt% urea, 25 wt% N,N,N′,N′-tetrakis(2-hydroxypropyl) ethylenediamine and 15 wt% Triton X-100.
5. Super glue (Loctite).
6. Optical projection tomography scanner Bioptonics 3001 (Bioptonics) or equivalent.
7. Image reconstruction software nRecon (Skyscan Pty Ltd., Kontich, Belgium).
8. 3D analysis software: open access, Fiji, Drishti, Vaa3D, and kidney analysis application [4]; commercial, Imaris (Bitplane AG), Amira (Visage Imaging Inc.), and Volocity (Perkin Elmer).

## 3 Methods

### 3.1 Whole-Mount Immunostaining

1. Dissect mouse embryonic kidney or genital ridges, in cold PBS.
2. Fix tissues in ice-cold 4% PFA for 30 min–1 h depending on mouse age and kidney size (E11.5 E13.5 30 min, E14.5–E16.5 60 min), at 4 °C.
3. Wash specimens twice for 10 min in PBS, to remove PFA.
4. These specimens are ready for whole-mount immunostaining (*see* **Note 1** for storing instructions).
5. Bleach specimens in 15% $H_2O_2$ and 15% DMSO diluted in absolute methanol at room temperature for 12 h while rocking.
6. Wash specimen in PBST-X-100 three times for 20 min while rocking.
7. Block the tissues in blocking solution containing 10% serum, 5% DMSO in PBST-X-100 (*see* **Note 2**). To ensure good blocking, use the serum from the same species where the secondary antibody was derived. Incubate for 12 h at room temperature while rocking.
8. Dilute primary antibodies in blocking solution, and incubate for 24 h at room temperature while rocking.
9. Wash in PBST-X-100 for 12 h; change to fresh PBST-X-100 for six times every 2 h.
10. Dilute secondary antibodies in blocking solution, and incubate for 24 h at room temperature while rocking.
11. Wash in PBST-X-100 for 12 h; change the solution for five times.
12. Proceed to preparation for OPT scanning (Subheading 3.4).

### 3.2 Whole-Mount In Situ Hybridization

We assume that the gene of interest was already subcloned to the vector (pBluescript II or others) for in situ probe preparation. The optimal DNA length for RNA probe production is about 800 bp +/−200 bp. It is important to prepare and run simultaneously antisense probes for positive signal and sense probes for negative control.

#### 3.2.1 Probe Synthesis

1. Mix reagents in the following order at room temperature: linearized plasmid 1 μg, 5× transcription buffer 4 μL, 10× DIG reaction mix 2 μL, 0.1 M DTT 2 μL, RNAse inhibitor 1 μL, polymerase 2 μL, DEPC water up to total reaction volume 20 μL.
2. Incubate for 2 h at 37 °C.
3. Remove 1 μL of the mixture and save for electrophoresis gel.
4. Add 2 μL DNase I to the rest of the probe synthesis reaction from **step 1**.
5. Incubate for 15 min at 37 °C.
6. Add 100 μL DEPC water, 10 μL 4 M LiCl, and 300 μL EtOH and mix. Incubate on dry ice or −70 °C for 20 min.
7. Spin in the tabletop centrifuge at 4 °C, at 12,000 × *g* for 10 min. Discard the supernatant.
8. Add 200 μL of 70% EtOH (ethanol diluted with DEPC water); spin in the tabletop centrifuge at 4 °C, at 12,000 × *g* for 10 min. Discard the supernatant.
9. Repeat **step 8**.
10. Invert the Eppendorf tube, and air-dry the pellet at room temperature for about 15 min.
11. Resuspend pellet in 45 μL DEPC water. Remove 1 μL of solution and save it for electrophoresis gel.
12. Add 45 μL formamide to sample, mix well, and store at −70 °C. Probes can be stored at −70 °C for several months without any damage.
13. Run electrophoresis gel with the samples from **steps 3** and **11**. Add 9 μL of 1× loading buffer to 1 μL of spared specimen, and load on the agarose gel. In the specimen from **step 3** will be visible both DNA and RNA bands, whereas in the specimen from the **step 11**, only RNA band should be visible. If RNA band is visible, proceed to in situ hybridization. If not, repeat and optimize probe synthesis.

#### 3.2.2 Whole-Mount In Situ Hybridization (WISH)

1. Dissect and fix embryonic kidney as indicated in Subheading 3.1, **items 1** and **2**. Wash specimens twice in PBS for 10 min, 50% ethanol for 10 min, and twice in 70% ethanol for 10 min. These specimens are ready for loading to in situ hybridization machine, or they can be stored at −20 °C for several months (*see* **Note 1**).

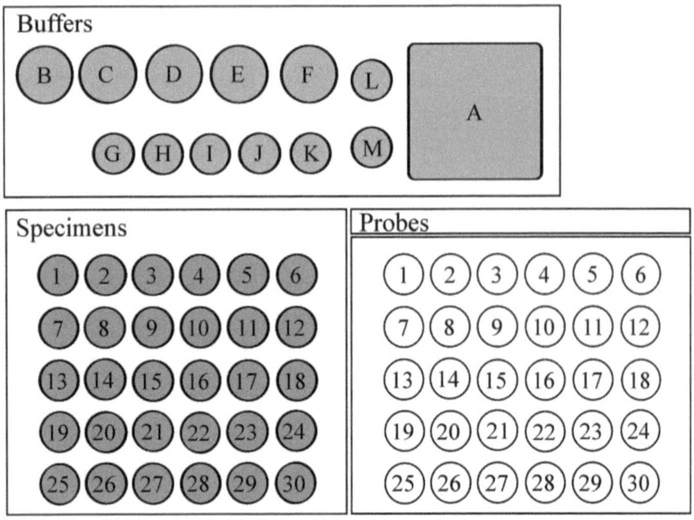


**Fig. 1** Schematic representation of buffer, specimen, and probe locations for automated WISH in InsituPro (Intavis AG Bioanalytical Instruments) device

2. Place specimens to InsituPro machine columns (comes with the device) filled with 70% ethanol. Use medium incubation columns, with outlet needles. If specimens are very small, use column inserts. Device can process 30 columns at a time and multiple specimens per column (Fig. 1) (*see* **Note 3** for manual WISH).
3. Prepare buffers for Day I (Table 1), and load them to the device in the order indicated in Fig. 1.
4. Cut the caps of from PCR tubes. Pipette 0.45 mL of pre-warmed hybridization buffer, and add 2.5 μL of RNA probe (about 0.1 μg/μL). The probe concentration might need to be optimized to get the best signal. Place tubes with probes in an order corresponding the specimen order (Fig. 1). Robotic arm will pipette preheated probe on the specimen during the first day. The hybridization temperature generally ranges from 60 to 70 °C. It might need to be optimized for each probe individually.
5. Set the program for InsituPro device to run following the manufacturer's instructions. An example for the program is presented in Table 2.
6. Based on this program, the device will stop twice. During the first stop, load buffers for Day II, and during the second stop, load buffers for Day III as in the order indicated in Table 1.
7. On Day III, once device has stopped, remove specimens from columns, and place to a 24-well plate.
8. Add 400 μL (or enough to cover the specimen) of BMP purple AP substrate precipitation solution. Wrap to aluminum foil to

**Table 1**
**The day-by-day work flow, buffer locations, and amounts for automated WISH adopted for InsituPro**

| | Buffers | | | |
|---|---|---|---|---|
| **Position** | **Day I** | **Day II** | **Day III** | **Buffer amount for 30 samples** |
| A | PBT-20 | MABT | | 100 mL/120 mL |
| B | EtOH 50% | | | 25 mL |
| C | Hybridization buffer | | | 25 mL |
| D | | Washing buffer I | | 35 mL |
| E | | Washing buffer II | | 35 mL |
| F | | MABT | | 35 mL |
| G | Proteinase K | | | 12 mL |
| H | 4% PFA | | | 12 mL |
| I | Glycine | | | 12 mL |
| J | 6% $H_2O_2$ | | | 12 mL |
| K | | | NTMT | 12 mL |
| L | | Blocking solution | | 12 mL |
| M | | Anti-Dig antibody | | 12 mL |

**Table 2**
**Detailed program description for automated WISH adopted for InsituPro (Intavis AG Bioanalytical Instruments) device**

| Nr. | Device action | Volume/ specimen | Time | Position/buffer | Comments |
|---|---|---|---|---|---|
| 1 | Set temperature regulation | OFF | | | |
| 2 | Rinse | 5000 μL | | | System wash |
| 3 | Aliquot | 200 μL | | B/ethanol | Rehydration |
| 4 | Wait | | 5 min | | |
| 5 | Incubate | 350 μL | 5 min | B/ethanol | Rehydration |
| 6 | Incubate | 200 μL | 0 min | | |
| 7 | Aliquot | 200 μL | | A/PBT-20 | Rehydration |
| 8 | Wait | | 5 min | | |
| 9 | Incubate | 350 μL | 15 min | A/PBT-20 | PBT-20 washes |
| 10 | Incubate | 350 μL | 15 min | A/PBT-20 | PBT-20 washes |

(continued)

**Table 2 (continued)**

| Nr. | Device action | Volume/ specimen | Time | Position/buffer | Comments |
|---|---|---|---|---|---|
| 11 | Incubate | 350 μL | 1 h | J/6% H2O2 | Bleaching |
| 12 | Incubate | 350 μL | 15 min | A/PBT-20 | PBT-20 washes |
| 13 | Incubate | 350 μL | 15 min | A/PBT-20 | PBT-20 washes |
| 14 | Incubate | 350 μL | 15 min | A/PBT-20 | PBT-20 washes |
| 15 | Incubate | 350 μL | 35 min | G/ProtK | Tissue permeabilization |
| 16 | Incubate | 350 μL | 5 min | I/glycine | Proteinase K inactivation |
| 17 | Incubate | 350 μL | 1 min | A/PBT-20 | PBT-20 washes |
| 18 | Incubate | 350 μL | 35 min | H/4% PFA | Post fixation |
| 19 | Incubate | 350 μL | 15 min | A/PBT-20 | PBT-20 washes |
| 20 | Incubate | 350 μL | 15 min | A/PBT-20 | PBT-20 washes |
| 21 | Incubate | 350 μL | 15 min | C/hybridization buffer | Wash |
| 22 | Set temperature regulation | T2 (HIGH) | | | Set hybridization temperature (60–70 C) |
| 23 | Incubate | 350 μL | 1 h | C/hybridization buffer | Pre-hybridization |
| 24 | Incubate | 350 μL | 15 h | Probes | Hybridization: probe will be pipetted on specimen |
| 25 | Wait for key | 350 μL | | | Device will stop; load Day II buffers and continue |
| 26 | Incubate | 350 μL | 30 min | D/washing buffer I | Post-hybridization washes I |
| 27 | Incubate | 350 μL | 30 min | D/washing buffer I | Post-hybridization washes I |
| 28 | Incubate | 350 μL | 30 min | D/washing buffer I | Post-hybridization washes I |
| 29 | Incubate | 350 μL | 30 min | E/washing buffer II | Post-hybridization washes II |
| 30 | Incubate | 350 μL | 30 min | E/washing buffer II | Post-hybridization washes II |
| 31 | Incubate | 350 μL | 30 min | E/washing buffer II | Post-hybridization washes II |
| 32 | Set temperature regulation | OFF | | | |

(continued)

**Table 2**
**(continued)**

| Nr. | Device action | Volume/ specimen | Time | Position/buffer | Comments |
|---|---|---|---|---|---|
| 33 | Wait | | 15 min | | |
| 34 | Incubate | 350 μL | 15 min | F/MABT | MABT washes |
| 35 | Incubate | 350 μL | 15 min | F/MABT | MABT washes |
| 36 | Incubate | 350 μL | 15 min | F/MABT | MABT washes |
| 37 | Incubate | 350 μL | 15 min | A/MBT | MABT washes |
| 38 | Incubate | 350 μL | 1 h | L/blocking solution | Blocking |
| 39 | Incubate | 350 μL | 6 h | M/antibody | anti-Dig antibody staining |
| 40 | Incubate | 350 μL | 15 min | A/MBT | MABT washes |
| 41 | Incubate | 350 μL | 15 min | A/MBT | MABT washes |
| 42 | Incubate | 350 μL | 15 min | A/MBT | MABT washes |
| 43 | Incubate | 350 μL | 20 min | A/MBT | MABT washes |
| 44 | Incubate | 350 μL | 20 min | A/MBT | MABT washes |
| 45 | Incubate | 350 μL | 20 min | A/MBT | MABT washes |
| 46 | Incubate | 350 μL | 40 min | A/MBT | MABT washes |
| 47 | Incubate | 350 μL | 40 min | A/MBT | MABT washes |
| 48 | Incubate | 350 μL | 40 min | A/MBT | MABT washes |
| 49 | Incubate | 350 μL | 50 min | A/MBT | MABT washes |
| 50 | Wait for key | | | | Device will stop; load Day III buffers and continue |
| 51 | Incubate | 350 μL | 15 min | K/NTMT | NTMT washes |
| 52 | Rinse | 350 μL | 5000 μL | | |
| 53 | Set temperature regulation | 350 μL | OFF | | End of program |

protect from light. Incubate at room temperature, and follow color developing every 30 min. If color does not develop, extend incubation overnight at +4C.

9. As soon as the color is strong and background remains clear, stop the reaction by washing in PBT-20 three times for 15 min.
10. Post-fix in 4% PFA for 1 h at room temperature.
11. Wash in PBS three times for 10 min.
12. Proceed to preparation for OPT scanning (3.4 Preparation for OPT scanning).

### 3.3 Combined Whole-Mount Immunostaining and Whole-Mount In Situ Hybridization

1. First, follow the whole-mount in situ hybridization protocol Subheading 3.2.2, after completion; wash specimens in PBS twice for 10 min.
2. Proceed with whole-mount immunostaining with antibodies of choice as indicated in Subheading 3.1.
3. Once both staining will be completed, proceed to preparation for OPT scanning in Subheading 3.4.

### 3.4 Preparation for OPT Scanning

We will refer to Bioptonics 3001 device; there are other open OPT platforms available [5–7], and the staining and preparation steps are equally suitable for all.

1. Observe the specimen in bright and fluorescence light under microscope, and remove all possible impurities adhered to the surface of the specimen.
2. Prepare 1% of regular or low melting point agarose dissolved in purified water and filtered (*see* **Note 4**).
3. Cut the tip of 5 mL syringe (Fig. 2a), secure the syringe to stand, put the plunger at 4 mL mark, pour 3 mL of the melted agar in (Fig. 2a), and put in the specimen (Fig. 2b). Keep on

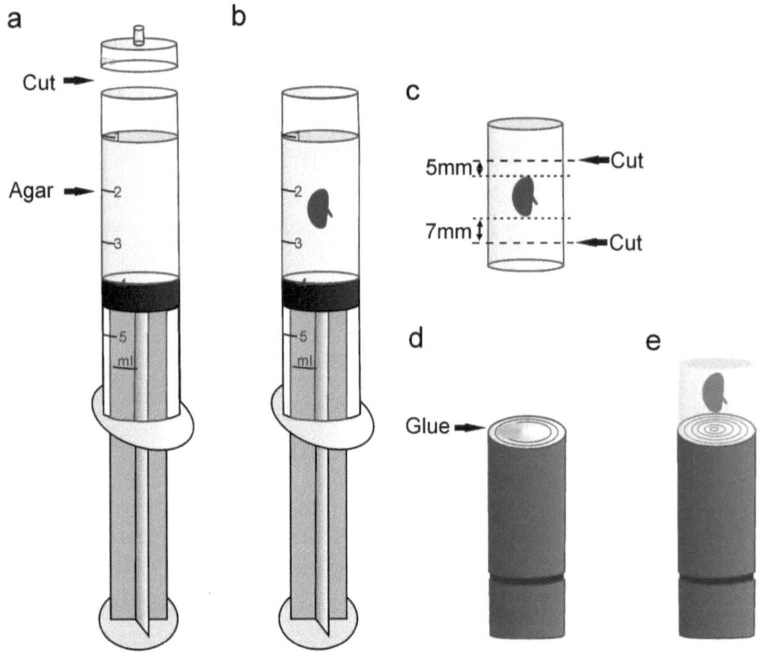


**Fig. 2** Tissue embedding to agar and mounting for OPT imaging. Cut the tip of syringe and pour agar (**a**). Place the specimen in, and move it around to position to the center as gel casts (**b**). Let the gel set, plunge the gel cylinder out from the syringe, and trim (**c**). Apply glue to the magnetic holder (**d**), and place the specimen (**e**)

swirling the specimen around with the plastic sticks or tweezers to position it to the center of the syringe as the gel is solidifying (Fig. 2b). Avoid bubbles. Let it set for 15 min at 4 °C.

4. Remove gel cylinder from syringe by plunging it out (Fig. 2c).
5. Trim gel cylinder by cutting excess gel. Leave 7 mm gap between specimen and bottom. This will ensure that there is enough working distance and the magnetic holder will not be visible during imaging. Leave 5 mm gap between specimen and the top cut (Fig. 2c).
6. Clear the specimen in BABB (Subheading 3.4, **item** 7) or ScaleCUBIC-1 (Subheading 3.4, **item 8**).
7. Benzyl alcohol: Benzyl benzoate solution (BA:BB, 1:2). Prior BABB clearance dehydrate tissue in 50% ethanol for 2 h, absolute ethanol 6 h, and fresh absolute ethanol for 6 h. Add BABB and incubate for 12 h. Change to fresh BABB after 12 h, and continue clearance for another 12 h or as long as specimen will become transparent (*see* **Notes 5** and **6**).
8. ScaleCUBIC-1 reagent clearance method will take longer time but will not require tissue dehydration. Change ScaleCUBIC-1 reagent after 24 h, and continue clearance until specimen will become translucent (transparent) (*see* **Note 5**).
9. Once specimen is cleared, remove it from the solution, and slightly dry the bottom side on a paper. Apply glue on the magnetic holder (Fig. 2d), and place the specimen on the top (Fig. 2e). Let it set for about 15 min.
10. Prepare OPT device for scanning following the appropriate instructions.
11. Place the prepared specimen into the device and scan it in clearance solution. Choose the correct acquisition channels: fluorescence and/or transmission. On Bioptonics 3001 device, set camera resolution to high (1024 × 1024), rotation steps to $0.45^{\circ}$, and the zoom to fit specimen to the field of view at all rotational angles. The specimen will rotate 360° and device will acquire 800 images.
12. Reconstruct scanned images to *z*-series to acquire OPT tomographic data by using nRecon software.
13. Proceed to analysis by your favorite software (software list is in Subheading 2.2, **item 8**).
14. An example of double immuno-labeled mouse kidney at E16.5 with anti-Troma (gray) and anti-Nephrin (red) antibodies (Fig. 3). The software-assisted quantification of nephrons is rather rapid and accurate (Fig. 4); in this case we have used spots detection feature available on Imaris software.

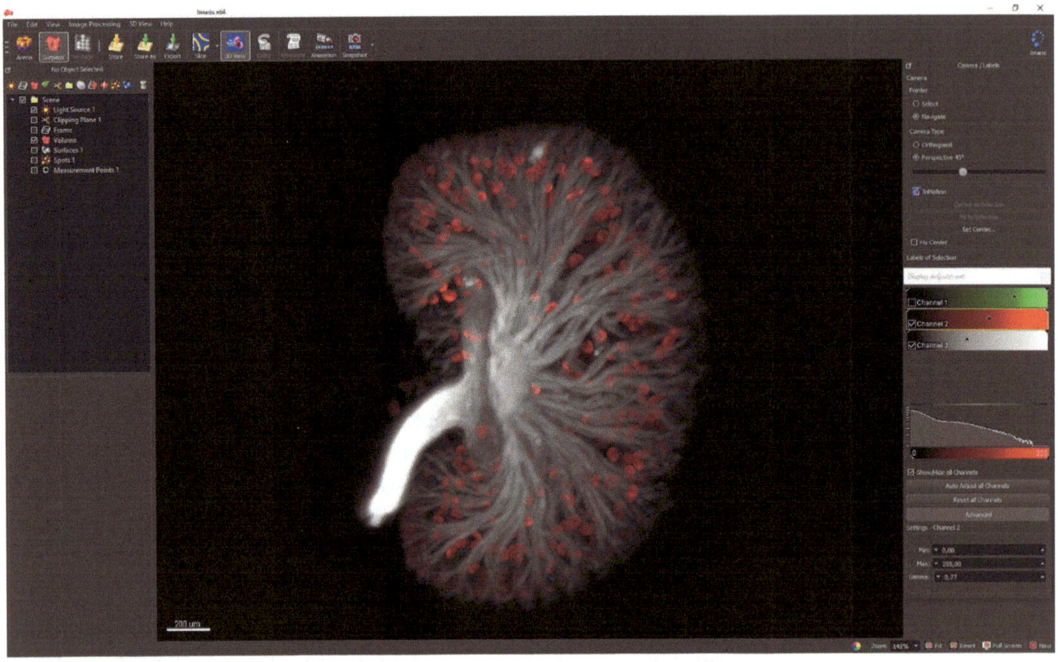

**Fig. 3** Representation of E16.5 mouse kidney double immuno-labeled with anti-Troma (gray) depicting ureteric tree and anti-Nephrin (red) antibodies showing nephrons. 3D visualization was done by Imaris program

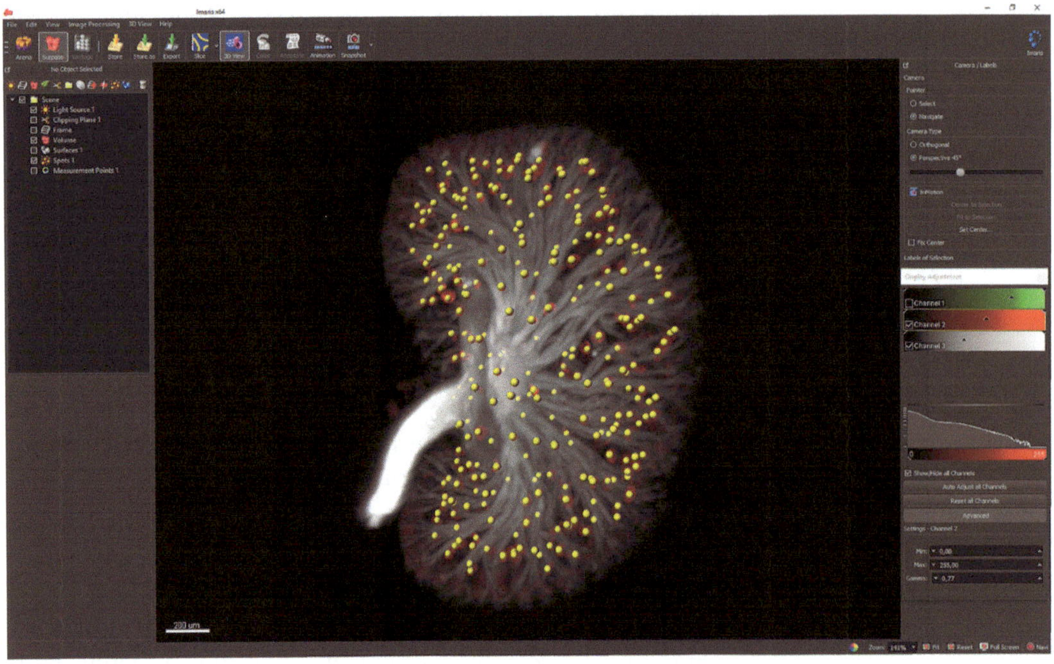

**Fig. 4** Example of automated nephron quantification by spot detection function (Imaris). Nephrons are depicted in yellow following nephrin staining (red)

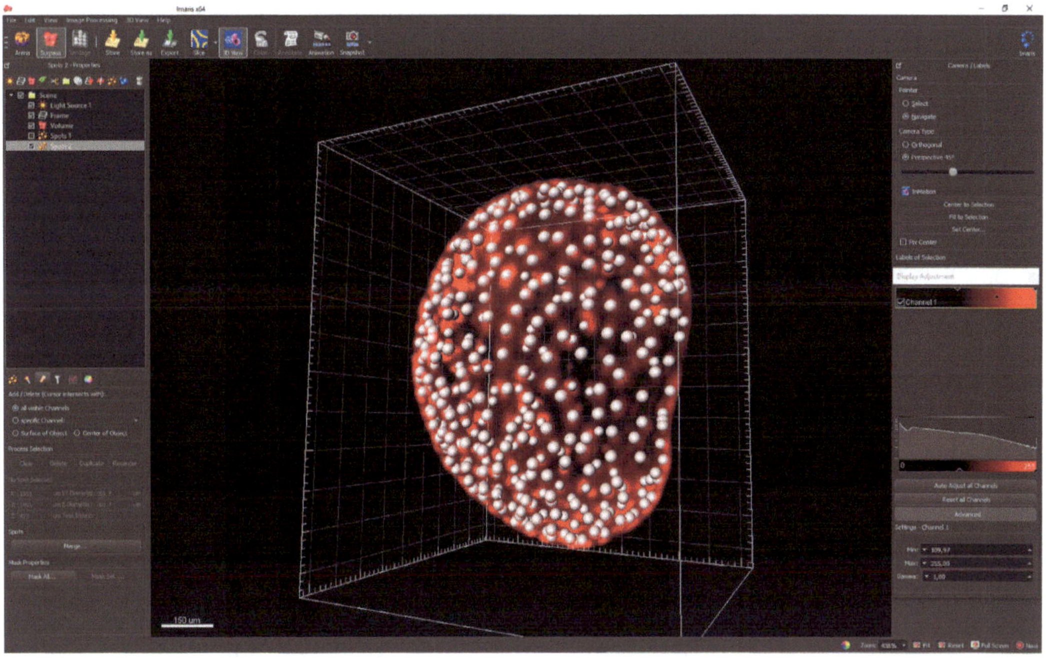


**Fig. 5** Ureteric bud (UB) tip quantification by spot detection function (Imaris). UB tips labeled with Wnt11 probe (red) by WISH in mouse kidney at E15.5

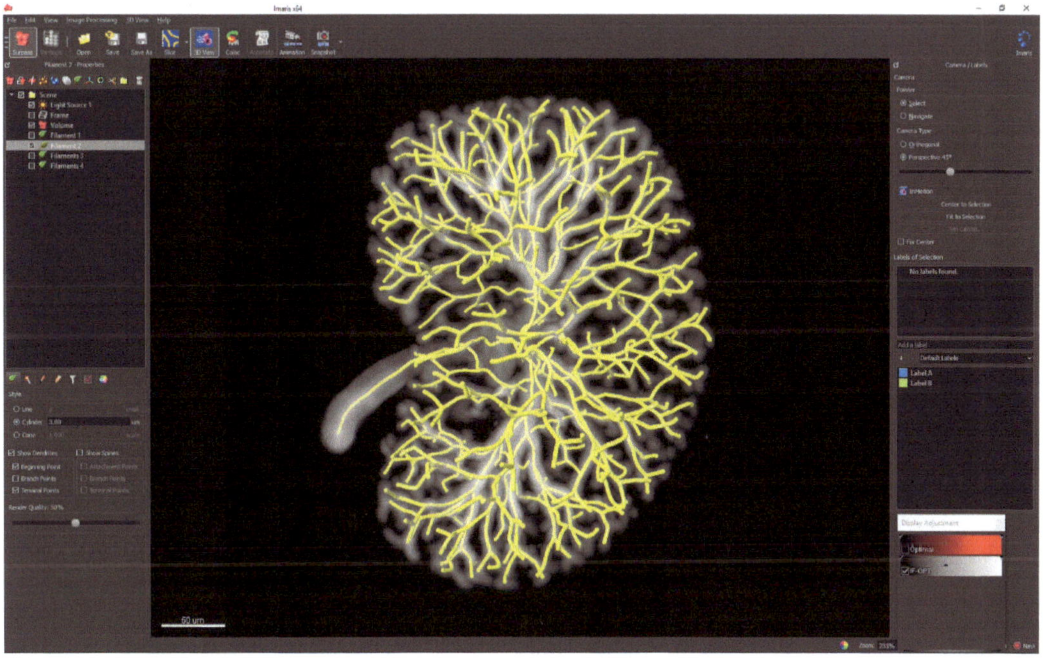


**Fig. 6** Ureteric tree analysis in developing mouse kidney at E15.5 by filament tracing function (Imaris). Kidney stained with anti-Troma antibody (gray)

15. An example of terminal tip quantification. We have used WISH method and RNR probe (*Wnt11,* depicted in red) to detect and quantify terminal tips of the developing ureteric tree in E15.5 mouse kidney and quantified them using spots detection feature available on Imaris software (Fig. 5).
16. An example of ureteric tree branching analysis of (in) developing mouse kidney at E15.5 by filament tracing function (Imaris) (Fig. 6). Analysis permits to acquire multiple parameters such as branching angles, number of terminal tips, branch length, and others. The kidney was whole-mount immunostained with anti-Troma antibody (gray).

## 4 Notes

1. For storing, dehydrate specimens in 50% ethanol once for 10 min and twice in 70% ethanol for 10 min. Tissues can be stored in 70% ethanol at −20 °C for several months. If specimens were stored in −70% ethanol, rehydrate them via ethanol series (70%, 50%, and 30%) to PBS before starting immunostaining.
2. In order to prevent bacteria growth during long incubation time, supplement buffers with 0.01% sodium azide.
3. The same protocol can be used to do WISH manually. Place specimens to 24-well plates for washing steps, and move them to glass bottles or 2 mL Eppendorf tubes for pre-hybridization, hybridization, and post-hybridization. Prepare all buffers as indicated, and follow the procedure and incubation times indicated in Table 2.
4. Use low melting point (LMP) agarose in cases when it is planned to reuse specimen after scanning for other applications. Otherwise, regular agarose can be used. In order to remove specimen from LMP agarose, rehydrate it from BABB to absolute ethanol for 6 h, change to fresh ethanol for 6 h, and place to 50% ethanol for 6 h and PBS for 6 h. Put agarose block in PBS and warm up in water bath at 60 °C. Once the gel softens, gently retrieve the specimen.
5. The BABB clearance is rather rapid and suits very well for immunostained specimens. In case of whole-mount in situ hybridization, BABB will gradually quench the color. The color will remain intact for about a week in BABB. In case of ScaleCUBIC-1, quenching is not an issue, but clearance will take longer time.
6. When working with BABB solution, use appropriate protection, and handle solution in the fume hood. Use glassware as BABB is toxic and dissolves (melts) plastic.

## References


1. Sharpe J, Ahlgren U, Perry P et al (2002) Optical projection tomography as a tool for 3D microscopy and gene expression studies. Science 296:541–545
2. Wilkinson D (1998) In situ hybridization: a practical approach (practical approach series). Oxford University Press, Oxford
3. Susaki E, Tainaka K, Perrin D et al (2014) Whole-brain imaging with single-cell resolution using chemical cocktails and computational analysis. Cell 157:726–739
4. Short K, Hodson M, Smyth I (2010) Tomographic quantification of branching morphogenesis and renal development. Kidney Int 77:1132–1139
5. Watson T, Andrews N, Davis S et al (2017) OPTiM: Optical projection tomography integrated microscope using open-source hardware and software. PLoS One 12:e0180309
6. Wong M, Dazai J, Walls J et al (2013) Design and implementation of a custom built optical projection tomography system. PLoS One 8: e73491
7. Gualda E, Vale T, Almada P et al (2013) OpenSpinMicroscopy: an open-source integrated microscopy platform. Nat Methods 10:599–600

# Part VI

## Drosophila as a Model to Target Conserved Renal Ontogenesis Genes such as the Teashirt

# Chapter 17

# Epithelial Function in the *Drosophila* Malpighian Tubule: An In Vivo Renal Model

**Shireen-A. Davies, Pablo Cabrero, Richard Marley, Guillermo Martinez Corrales, Saurav Ghimire, Anthony J. Dornan, and Julian A. T. Dow**

## Abstract

The insect renal (Malpighian) tubule has long been a model system for the study of fluid secretion and its neurohormonal control, as well as studies on ion transport mechanisms. To extend these studies beyond the boundaries of classical physiology, a molecular genetic approach together with the 'omics technologies is required. To achieve this in any vertebrate transporting epithelium remains a daunting task, as the genetic tools available are still relatively unsophisticated. *Drosophila melanogaster*, however, is an outstanding model organism for molecular genetics. Here we describe a technique for fluid secretion assays in the *D. melanogaster* equivalent of the kidney nephron. The development of this first physiological assay for a *Drosophila* epithelium, allowing combined approaches of integrative physiology and functional genomics, has now provided biologists with an entirely new model system, the *Drosophila* Malpighian tubule, which is utilized in multiple fields as diverse as kidney disease research and development of new modes of pest insect control.

**Key words** *Drosophila melanogaster*, Epithelia, Renal, Malpighian tubules, Fluid secretion assays, Kidney disease model, Functional genomics

## 1 Introduction

### *1.1* D. Melanogaster *for Functional Genomics and as Human Disease Model*

The discovery of the *Drosophila melanogaster white* mutation by Thomas Hunt Morgan [1] and the award of six Nobel prizes to date (Table 1) have firmly established *D. melanogaster* as a key genetic model for developmental biology, neurobiology, innate immunity, and chronobiology.

*D. melanogaster* is itself an unremarkable, but nonetheless highly representative, dipteran insect. The special qualities which it brings to research questions are largely the result of insightful scientific exploitation over a century. Its short life cycle, fecundity, and ease of rearing led it to be selected as a genetic model in 1910 by Morgan, Bridges, Sturtevant, and Muller [2]. The discovery of

Seppo Vainio (ed.), *Kidney Organogenesis: Methods and Protocols*, Methods in Molecular Biology, vol. 1926, https://doi.org/10.1007/978-1-4939-9021-4_17, 

**Table 1**
**Nobel prizes awarded for multiple research areas using *D. melanogaster***

| Year | Awardees | Field |
|---|---|---|
| 1933 | Thomas Hunt Morgan | Chromosomes in heredity |
| 1946 | Hermann Joseph Muller | X-ray irradiation and mutation rates |
| 1995 | Edward B Lewis, Christiane Nüsslein-Volhard, and Eric F Wieschaus | Genetic control of embryonic development |
| 2004 | Richard Axel | Odor receptors and the organization of the olfactory system |
| 2011 | Jules A Hoffmann | Innate immunity |
| 2017 | Jeffrey C Hall, Michael Rosbash, and Michael W Young | Molecular mechanisms of circadian rhythms |

mutants, and of predictable patterns of recombination between them, led to mutant stocks being maintained in stock centers for future use. Development of powerful resources over the last three to four decades, including P-elements for both transgenesis and mutagenesis [3], the powerful GAL4/UAS binary expression system (*see* below) [4] and luminescent and fluorescent reporters for cytoplasmic and organellar calcium [5–7], cAMP [8], and chloride [9], and use of optogenetic tools [10, 11], has allowed researchers to address fundamental scientific questions using increasingly sophisticated approaches. Finally, the discovery that RNA interference (RNAi) was capable of knocking down gene expression relatively selectively allowed the possibility that comprehensive mutant resources would be generated for *Drosophila*. Thus, there are now transgenic fly lines with RNAi constructs directed against nearly every gene in the genome [12].

The gene expression resources for *D. melanogaster*, including FlyBase [13] but particularly the first tissue-specific gene expression atlases for any organism—FlyAtlas.org [14]—and its subsequent updates to include RNAseq and miRNA data [15] have been exemplars of community resources which have galvanized research in *D. melanogaster* for a new generation of researchers. Together with the resources for genetics and physiology, these ‘omics resources have also allowed development of *D. melanogaster* as a key model, linking gene function with physiology (“integrative physiology”), particularly for epithelial function [16, 17].

Although Drosophila and human are evolutionary divergent by 400 million years, ~70% of *Drosophila* genes have human disease orthologues; and although humans and flies differ in morphological features and molecular mechanisms of development, cellular and physiological processes are highly conserved. Furthermore, several *Drosophila* internal organ systems are functionally analogous to human systems. Altogether, *D. melanogaster* is now an accepted model for many human diseases including those of the cardiovascular and nervous systems and kidney [18, 19]. As such, there is now increasing use of *Drosophila* as a highly effective phenotypic, in vivo screening tool for therapeutics and also toward understanding disease mechanisms [20–22].

### 1.2 Osmoregulation and Drosophila Malpighian Tubules

Insects, by virtue of their small size, have a large surface-to-volume ratio and so are under continual osmotic stress. For terrestrial insects, defense against desiccation is provided by behavioral, physiological, and physical means. In particular, an impermeable cuticle limits evaporative water loss, while the excretory system (composed of Malpighian tubules and hindgut) dynamically balances primary urine generation and secondary reabsorption, so achieving a compromise between osmoregulation and excretion [23–25].

Malpighian tubules were identified and so named after Marcello Malpighi, who first identified these tissues in the silkworm, *Bombyx mori* [26]. Malpighian tubules, in all insects which have them, are fluid-secreting epithelia [25], and tubules of *D. melanogaster* are known to be the fastest fluid-transporting epithelia known in biology [27] (Fig. 1).

The *D. melanogaster* kidney is formed of four Malpighian tubules, analogous to a single vertebrate nephron [28]. Extensive work on the development of *D. melanogaster* tubules [29] has demonstrated developmental events including signaling pathways, as well as key mechanisms in tubule development, e.g., mesenchymal-to-epithelial transition, that are conserved in mammalian systems [30].

Tubules emerge from the hindgut, just behind the junction with the midgut, and constitute a pair of anterior and pair of posterior tubules with a common ureter [30]. Each tubule is composed of ~150 cells [31] and can be divided into four domains based on morphology and function: initial segment, transitional segment, the fluid-secreting main segment, and lower tubules [31, 32]. Additionally, the initial segment in anterior tubules is enlarged and contains calcium concretions toward whole body calcium homeostasis [6, 33, 34]. Primary urine formation in tubules is due to active transport processes and is formed initially by the main segment of both anterior and posterior tubules as K+-rich fluid [32]. The action of the lower tubule segment is to reabsorb KCl and fluid secreted by the main segment, before this enters the hindgut for final formation of urine [35].

*Drosophila* tubule ion transport pathways have been extensively mapped [30] and occur within two major cell types, the principal and stellate cells [36], which allow functional separation of ion transport and cell signaling pathways [16] for physiological function. Principal cells contain the large ion transport complexes, e.g., the vacuolar $H^+$-ATPase [37, 38] and the $Na^+/K^+$ exchanger

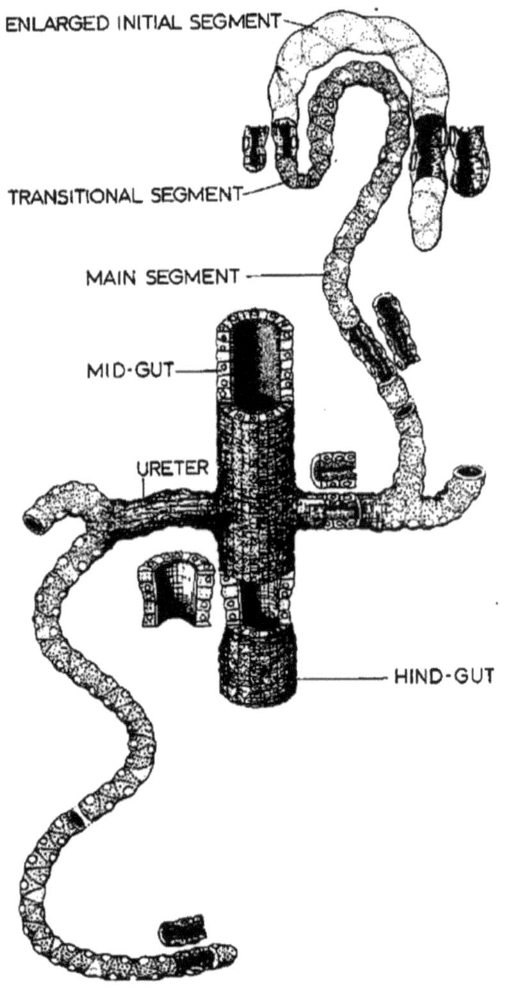


**Fig. 1** Classical morphology of the *D. melanogaster* Malpighian tubules showing one each of the anterior tubule pair (top) with indicated regions including the main (fluid-secreting) segment and enlarged initial segment and one of the posterior tubule pair (bottom) [Reproduced with permission from [28]]

[39]. Stellate cells, by contrast, express water channels (aquaporins) [40] and control chloride flux via chloride channels including those of the CLC family [9], Fig. 2.

### 1.3 Malpighian Tubules as Key Tissues for Survival

It is now also known that insect tubules perform many more functions than osmoregulation. Tissue-specific transcriptomics of *D. melanogaster* [14, 42] have led the way in assigning novel functions to *D. melanogaster* tubules [36]. Moreover, the anterior and posterior tubules exhibit transcriptome and functional asymmetry [43], suggesting further intriguing possibilities of anterior- and posterior-specific roles for each pair of tubules. Assignment of novel functions such as detoxification and xenobiotic handling and

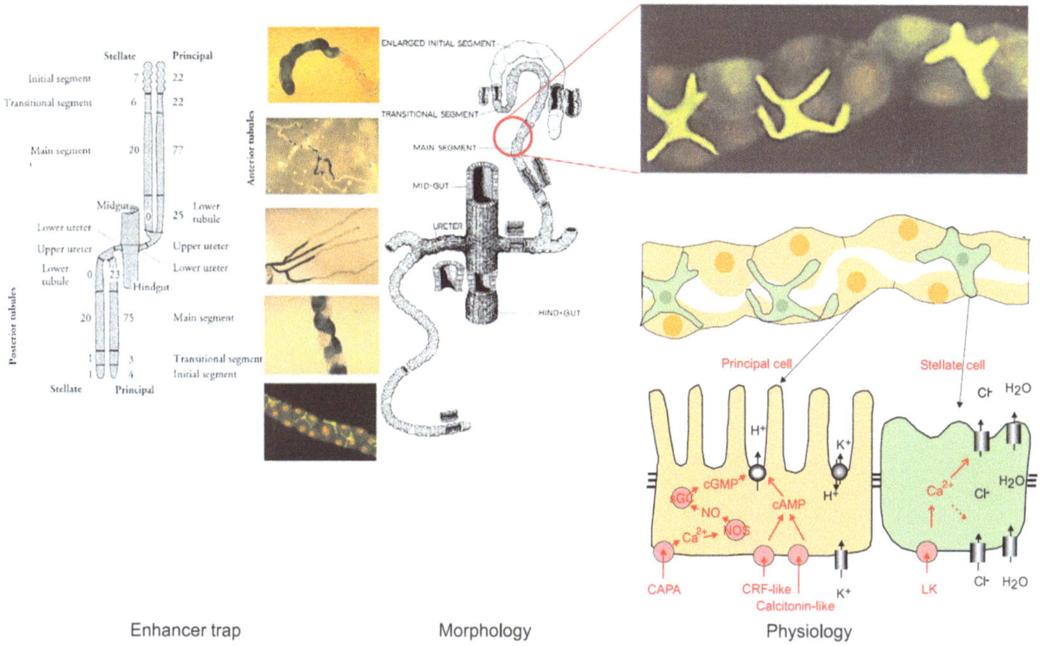


**Fig. 2** Three views of the *D. melanogaster* tubule. Center: the classical morphological view [28]. Left: a genetically-derived enhancer trap view, with some representative enhancer trap expression patterns [31]. Right: summary of the physiology of the tubule [16]; the stellate cells in the micrograph (top panel) are lit with UAS-GFP, driven by the stellate cell-specific GAL4 lines, c724. *CRF* corticotropin-releasing factor, *LK* leucokinin from [41]. Reproduced with permission from [36]

stress sensing of oxidative, osmotic (ionic/salt), and immune challenges by transcriptomic analysis has been underpinned by functional and physiological analyses [44, 45]. Thus, as with vertebrate kidneys, the tubules are mission-critical tissues for survival.

The cell- and tissue-specific function of genes and proteins in renal function can be assessed via the *D. melanogaster* GAL4/UAS binary expression system [4]. The GAL4/UAS targeted expression system is based on transgene expression directed by the binding of yeast transcription factor, GAL4 (for which there are no endogenous targets in the fly), to its upstream activation sequence (UAS), allowing expression of any transgene cloned downstream of the UAS element. Cell- and tissue-specific GAL4 lines ("driver" lines) are available for many cells and tissues in *D. melanogaster*, including the Malpighian tubules [31], allowing highly specific, targeted expression of transgenes in vivo (Fig. 3).

### 1.4 The Malpighian Tubule as a Model for Human Renal Disease

The rapid rate of fluid secretion by the tubule suggests that it is an excellent model for basic epithelial physiology, and the demonstration of its importance in xenobiotic handling [47, 48] makes this a tissue of great applied interest. However, there is also the potential to exploit similarities between the insect and human renal systems.

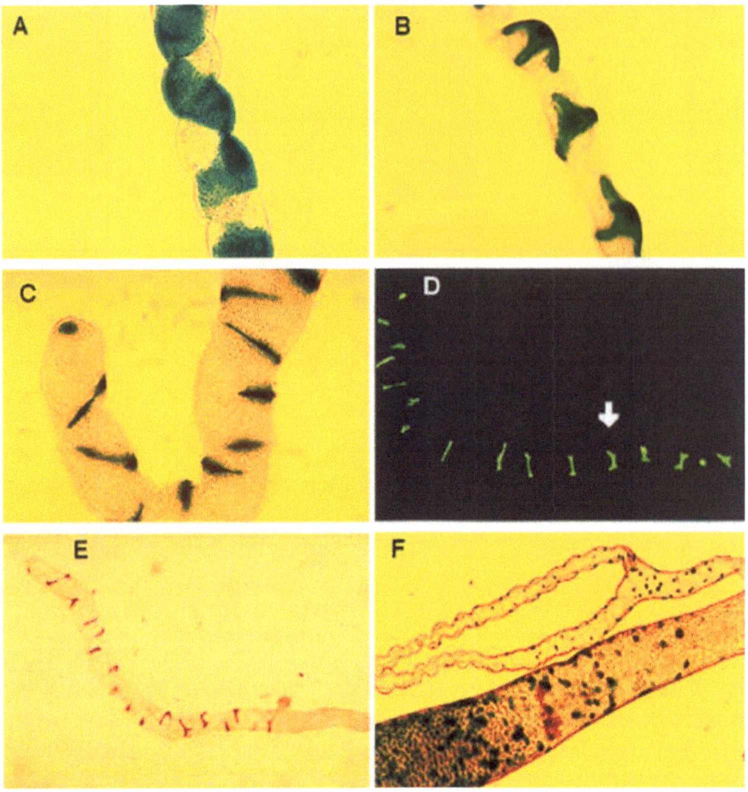


**Fig. 3** Genetic mapping of tubule cell types. (**A**) Lines c324 and c374 distinguish subsets of morphologically indistinguishable principal epithelial cells in the main segment. (**B**) Lines c710 and c724 are expressed uniquely in the stellate cells of adult and larvae. (**C**) Lines c710 and c724 also mark bar-shaped cells in the initial segment of anterior tubules. (**D**) Type II (stellate) cell morphology switches from bar-shaped to stellate cells at the transitional-main tubule boundary (arrow). The tip of the tubule is to the left. (**E**) Line c649 marks only the bar-shaped type II cells of anterior tubule initial and transitional segments. (**F**) Line c325 demonstrates the presence of a small, previously unreported cell type in the lower tubules and ureter and of homologues in the posterior midgut. Cell-specific lines, e.g., c324 and c710/724, have been used as GAL4 drivers. Reproduced with permission from [31]. New cell-specific tubule GAL4 lines have also since been developed allowing precise targeting of genes of interest to either principal or stellate cells [9, 46]

Across 400 million years of divergent evolution, one might expect very little in common between human and fly renal tubules—however, as in mammalian kidney, the $Na^+$, $K^+$-ATPase plays a major role in insect tubule function [39, 49–51].

Moreover, the discovery of a *Drosophila* Malpighian tubule phenotype in the first animal knockout of a V-ATPase [52] presaged the identification of a human renal phenotype arising from the first human V-ATPase mutation to be discovered [53]. Since then it has been demonstrated comprehensively in *Drosophila* and to a lesser extent in mammalian systems that mutations in any of the epithelial V-ATPase subunits can result in renal phenotypes similar to that first described with the V-ATPase null [38].

However, is the insect tubule a good model for diseases other than tubular acidosis? There are 13,500 genes in *Drosophila* and a larger number in humans. Work based on transcriptomics [42] produced a list of tubule-specific (or at least enriched) genes with a high degree of homology to human disease genes. In keeping with the argument that human and fly renal systems demonstrate a high degree of functional and genetic similarity, the list was also enriched for classical human renal disease loci [42] including Bartter's syndrome, xanthinuria type 1, and hyperoxaluria type 1, implying far greater commonality between the two systems than previously supposed.

Since this discovery, functional studies including those utilizing new technologies, e.g., metabolomics [54], have been performed to support these observations and have proceeded to validate the *Drosophila* tubule model for kidney disease [19].

### 1.5 Functional Analysis of Tubule Function

Analysis of tubule function and physiology includes measurement of fluid secretion rates [32], ion and ion transport measurement via electrophysiology [55], and targeted, in vivo reporter constructs, e.g., those used to measure intracellular calcium [6]. The impact of alterations in gene expression in specified cells of the MTs on the whole organism can also be assessed via organismal survival assays including under stress conditions, e.g., as for genes encoding the CLC family of chloride channels [9].

Here, we focus on measurement of fluid secretion rates as functional, physiological outputs of the Malpighian tubule which can be used to determine not only the integrative physiology of epithelial function but also to fill the "phenotype gap" for functional genomics [56].

Fluid secretion assays can also be used to comprehensively screen for gene-specific effects and also act as an organotypic, in vivo screen for therapeutics.

## 2 Materials

### 2.1 Drosophila Melanogaster

Wild-type strains of *D. melanogaster* include Oregon R and Canton Special and have been maintained in laboratories since 1925 [57]. Either strain can be used but whichever is used should be standardized across all experiments.

Oregon R flies were raised on standard fly medium (Table 2) in tubes at 23 °C and in ambient humidity with a 12:12 h L/D cycle [84].

Fluid secretion assays can be performed on Malpighian tubules at different stages of the *Drosophila* life cycle including mature

**Table 2**
**Ingredients for *Drosophila* standard medium**

| Standard fly food | Amounts of reagent per 10 L |
|---|---|
| Agar | 100 g |
| Sucrose | 150 g |
| Glucose | 300 g |
| Yeast | 350 g |
| Maize meal | 150 g |
| Wheat Germ | 100 g |
| Treacle | 300 g |
| Soya Flour | 10 tbsp |
| Nipagen | 100 mL |
| Propionic Acid | 50 mL |

third-instar larvae and adults at 1 week postemergence. Although no consistent age effects were observed at either of these stages, the tubules of younger adults (4–7 days after eclosion) appeared to be less easily damaged by dissection. Although no obvious sex difference in tubular secretion was observed, only female adults were selected, partly in order to control for undetected differences and partly because their tubules are larger than those from the smaller males.

### 2.2 Solutions

Schneider's insect culture medium (Thermo Fisher) was used for dissection of intact tubules to maintain the viability of the isolated tissue for as long as possible. The composition of Schneider's medium is 3.33 mM glycine, 2.3 mM L-arginine, 3.01 mM L-aspartic acid, 0.496 mM L-cysteine, 0.417 mM L-cysteine, 5.44 mM L-glutamic acid, 12.33 mM L-glutamine, 2.58 mM L-histidine, 1.15 mM L-isoleucine, 1.15 mM L-leucine, 9.02 mM L-lysine hydrochloride, 5.37 mM L-methionine, 0.909 mM L-phenylalanine, 14.78 mM L-proline, 2.38 mM L-serine, 2.94 mM L-threonine, 0.49 mM L-tryptophan, 2.76 mM L-tyrosine, 2.56 mM L-valine, 5.62 mM β-alanine, 5.41 mM $CaCl_2$, 15.06 mM $MgSO_4$, 21.33 mM KCl, 3.31 mM $KH_2PO_4$, 4.76 mM $NaHCO_3$, 36.21 mM NaCl, 4.94 mM$Na_2HPO_4$, 1.37 mM α-ketoglutaric acid, 11.11 mM D-glucose, 0.862 mM fumaric acid, 0.746 mM malic acid, 0.847 mM succinic acid, 5.85 mM trehalose, and 2000 mg/L yeast plate (http://www.thermofisher.com/uk/en/home/technical-resources/media-formulation.124.html).

For the secretion assays, a 1:1 (v/v) mixture of Schneider's insect culture medium (Thermo Fisher) and *D. melanogaster* saline,

**Table 3**
**Components of standard *Drosophila* saline [32]**

| Compound | mmol $L^{-1}$ |
|---|---|
| NaCl | 117.5 |
| KCl | 20 |
| $MgCl_2$ | 25 |
| $CaCl_2$ | 2 |
| $NaHCO_3$ | 10.2 |
| $NaH_2PO_4$ | 4.5 |
| HEPES | 8.6 |
| Glucose | 20 |

pH 7.0 (Table 3), was used, where secretion was faster and maintained for longer in a 1:1 mixture of standard saline with Schneider's medium. Osmolarity of the solution is critical to achieve robust, consistent secretion rates; osmolarity of *Drosophila* saline is 300 Osm/kg. To avoid calcium salt precipitation, the pH was adjusted to 7.00, and $NaHCO_3$ and $NaH_2PO_4$ were each weighed separately and added to the solution just before the experiment, after which freshly weighed glucose was added.

### 2.3 Fluid Secretion Assay

Secretion assay plate: The following materials are required to create a secretion assay plate: 55 mm petri dish, paraffin wax (Sigma-Aldrich/327212), 2 mm drill bit, nonmagnetic steel pins, e.g., entomological "minuten" pins, and mineral oil (Sigma-Aldrich/330760, heavy viscosity) although ideally a mix of heavy and light viscosity grades so the fluid droplets (Fig. 5b) are neutrally buoyant and do not drift out of focus.

Tubule dissection: No. 5 Biology Grade dissection forceps are used, and a 3 mm diameter glass rod pulled to a fine (100 μm tip) is required to transfer the tubules into the wells.

Amaranth dye (Sigma-Aldrich/A1016), Schneider's insect culture medium (Thermo Fisher), and *D. melanogaster* saline, pH 7.0 (Table 3), are also required.

### 2.4 Microscopy for Visualization of Fluid Secretion Assays

A stereo dissecting microscope with a stage micrometer is required for fluid secretion assays, and this should be configured for a specified (fixed) magnification. An appropriate conversion should also be factored in for the graticule that is used to measure the size of the secreted bubble (Fig. 5b). For our use, the stereo dissecting microscope is equipped with standard magnification of 6–50× and eyepieces 10×/21, with a graticule (12 mm/120 div or 14 mm/140div) in one of the eyepieces.

## 3 Methods

### 3.1 Setup of Secretion Assay Plate

A 55 mm petri dish was filled with a 7 mm layer of hot paraffin wax and left to cool. Twenty-one wells distributed across three lanes were carved using a 2 mm drill bit. An entomological minute pin was inserted beside each well, at an empirically determined distance for the species under study. The dish was then covered with mineral oil. A 1:1 (v/v) mixture of Schneider's insect culture medium (Thermo Fisher) and *D. melanogaster* saline, pH 7.0 (Table 3), was prepared and a small volume of amaranth added for contrast purposes. A 10 μL droplet of this bathing medium was deposited in each well.

### 3.2 Tubule Dissections

For the adult fly, unless otherwise stated/required, flies were aged for 7–10 days at 22–25 °C prior to dissection. Flies were briefly anesthetized on ice prior to transfer into a dissecting dish containing Schneider's liquid medium. Using No. 5 Biology Grade dissection forceps, each animal was immobilized by lightly grasping the junction between the thorax and abdomen and firmly gripping the external genital plate. The internal genitalia, mid- and hindgut, and Malpighian tubules were then drawn out of the abdomen by a "clean pull" of the forceps holding the genital plate. The tubules could then be isolated from the extraneous tissues by cutting the junction of the ureter and guts with the forceps. Dissected tubules could then be transferred to the secretion assay plate wells using a fine glass rod (Fig. 4).

Dissections for larvae are essentially similar in nature. Larvae at the appropriate developmental stage (e.g., L3 molt) are selected and washed in ice-cold Schneider's to remove extraneous food prior to transfer to a dissecting plate with more Schneider's medium. An opening is made in the posterior end of the larva, the mouthparts are held with a fine dissecting forceps, and the cuticle is then rolled over the forceps. The anterior and posterior

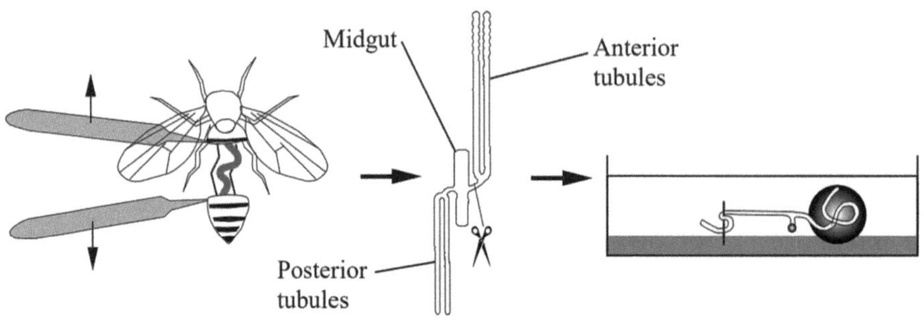


**Fig. 4** Overview of the dissection and preparation of intact Malpighian tubules for fluid secretion assays. Illustration of dissection of an adult *D. melanogaster* fly, isolation of intact tubules, and setup of tubules in the assay system. Reproduced with permission from [32]

ends of the larvae are then gripped by forceps and gently pulled, effectively inverting the whole larvae, exposing the internal organs/tissues while leaving them attached in situ. The Malpighian tubules may then be isolated from the surrounding tissues and then transferred to wells of the secretion assay plate using a fine glass rod.

### 3.3 The Assay System

The *Drosophila* Malpighian tubule fluid secretion assay [32] was adapted from the Ramsay assay [58] classically employed for insect tubules [85]. This has allowed the application of classical techniques even to these tubules from a genetic model organism, the smallest yet studied. For *D. melanogaster* tubules, a pair of tubules (*see* **Note 1**)—still linked by their attachment to the remaining part of the common ureter—is placed in each 10 μL drop of bathing medium above, in each well of the secretion assay plate (Section 3.1, Fig. 5a). One of the pair of tubules is pulled out of the drop and wrapped around the steel pin, to which it adheres by surface tension (Figs. 4 and 5b). The use of one tubule of each pair as an anchor for the other allows virtually the entire main segment, secretory region of a tubule to be bathed in the saline drop. The lower segment (Fig. 1) is the reabsorptive region [35] and should not be bathed in the saline drop. The secreted fluid emerges from the aperture at the cut end of the common ureter. Droplets of secreted fluid are removed at intervals of 10 min after a measurement at time zero with a fine glass rod and suspended in the paraffin oil. Measurement of the diameter of the spherical droplet with an ocular micrometer (ideally measured twice, orthogonally) allows calculation of the volume of fluid secreted ($4/3\pi r^3$), and measurement of the time interval in which the drop was secreted allows the secretion rate to be calculated (volume/time) expressed as nL/min

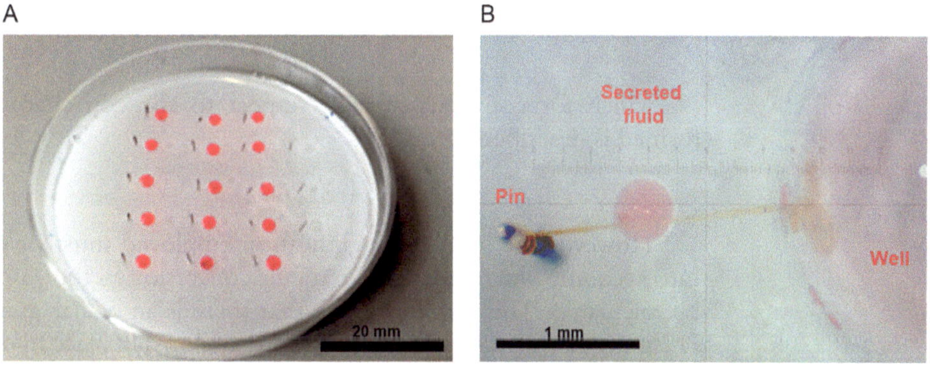


**Fig. 5** Assay setup. (**A**) An experimental assay plate showing parallel rows of wells containing 10 μL of bathing medium with amaranth, each well with a steel pin alongside. For clarity, the plate shown allows 15 fluid secretion measurements with a pair of tubules per well. In practice, most assay plates are set up to contain 21 wells allowing secretion rate measurements of up to 21 tubule pairs per experiment. (**B**) Magnification of one well/tubule pair in the secretion assay plate showing micrometer, pin, and well and also indicating the fluid secretion bubble emerging from the cut ureter under the paraffin oil

(*see* below). To capture the dynamic changes in fluid secretion, especially after neuroendocrine stimulation, we sample as often as possible for a period of up to 60 min (*see* **Note 2**); in practice, the smallest drop that is easily removed (around 0.2 nL) limits the interval to around 10 min, and measurements at intervals higher than 20 min are not recommended. However, the assay is intrinsically parallel; and with practice, secretion rates of 20 tubules can be measured simultaneously in a single dish at intervals of 10 min. Additionally, as the small size of tubules allows up to 20 tubules to be assayed simultaneously in a single dish, this permits good internal controls for extraneous variables.

No difference in secretion rate was observed between anterior and posterior tubules in either larvae or adults, and, accordingly, both were used in subsequent experiments. Three morphologically distinct regions are distinguishable (at both light and electron microscopic levels) in the anterior tubules and two in the posterior tubules of *Drosophila* [28]. The possibility that fluid secretion might be restricted to tubule subregions was studied in a series of experiments, in which varying lengths of tubule were allowed to remain in the drop [35]. The most distal (white) region of the anterior tubules failed to secrete measurable volumes of fluid over several hours; the main central segment of both anterior and posterior tubules secreted fluid; in both, the lower tubule and collecting duct showed weak reabsorptive activity [35]. Provided that the entire main segment was included in the bathing drop, variation in the precise overall length of tubule included did not therefore introduce serious variability in the measured readings.

### 3.4 Calculation of Fluid Secretion Rate

The *general formula* for measuring the volume of the secretion "bubble" with the first assumption being that this is a perfect sphere:

$$V = 4/3\pi r^3$$

which for convenience with respect to the conversion performed is described as:

$$V = 4/3\pi (D/2)^3$$

However the level of magnification employed must be taken into account, which requires a graticule conversion factor (GCF) to be calculated. This requires firstly that there is an ocular graticule (12 mm/120 divisions or 14 mm/140 divisions) in place in one of the eyepieces and a stage micrometer. Using the stage micrometer as a reference, the distance each division is worth at × 50 magnification is calculated in mm. In our microscopy system, 1 division equals 0.02 mm; ergo the GCF is 0.02.

So the calculation then becomes:

$$V = 4/3\pi(D/2 \times \mathrm{GCF})^3$$

The second assumption is that 1 $mm^3$ = 1 **μL**. Since the secretion rate is in the nanoliter range, this is multiplied by 1000 to transform the result to nanoliters: $V = 4/3\pi(D/2 \times \mathrm{GCF})^3 \times 1000$.

This is then divided by the measured time interval in minutes (T1-T0), typically 10 min. So the final calculation is:

$$V = 4/3\pi(D/2 \times \mathrm{GCF})^3 \times 1000/(T_1 - T_0)$$

$$= \text{secreted fluid in nL/min.}$$

### 3.5 Statistics

Mean and SEM secretion rates are calculated every 10 min. To calculate %Δ increase in fluid secretion compared to basal or control levels, the following formula is applied to each secreting tubule: [(Maximum response after stimulation—mean basal response before stimulation)/(mean basal response before stimulation)] ×100.

### 3.6 Fluid Secretion Measurements to Determine Gene Function

Direct measurement of fluid secretion by tubules of mutant and/or transgenic *Drosophila* with either targeted expression of RNAi (loss-of-function) or overexpression (gain-of-function) constructs has provided key insights into gene function in Malpighian tubules. These functional studies have also supported previous work that has delineated gene and "core motif" specification of the tubules as transporting epithelia [42, 59] including that incorporating other approaches, e.g., genetics [60], pharmacology [39, 55, 61–65], and electrophysiology [66–68]. Examples of some of the functional studies of fluid secretion in gene mutants and/or transgenic lines are provided in Table 4. Characterized genes have to date included those encoding transcription factors, transporters, cell signaling components, neuropeptides, and neuropeptide receptors (G-protein-coupled receptors).

## 4 Notes

1. Both larval and adult tubules can be used for secretion assays; however, adult tubules have proved more robust and so have been used since for definitive studies. The results of a typical experiment with adult tubules are shown in Fig. 6.
2. Unstimulated tubules secreted fluid steadily for over 5 h [32]. Experiments recording the average/basal rate of secretion show that tubules can continue to secrete for up to 15 h, unusual even in insect transporting systems.

**Table 4**
**Determination of gene function by physiological secretion assays**

| Gene | Function | References |
|---|---|---|
| Fasciclin2 | Regulates brush border length, microvilli size, shape and organization | [69] |
| Irk1 and Irk 2 | Play (redundant) roles in transepithelial $K^+$ transport and fluid secretion | [70] |
| Chloride channel<br>CLC-a in stellate cells and CLC-c in principal cells | Each CLC is required for stimulated fluid secretion rates | [9] |
| $Na^+$-$K^+$-$2Cl^-$(NKCC) cotransporter Ncc69 | Modulates stimulated fluid secretion rates | [71] |
| Nha1 | Required for basal fluid secretion rates | [72] |
| Vha55 alleles | Required for stimulated fluid secretion rates | [60] |
| Adenine nucleotide translocase (ANT, the mitochondrial inner membrane ATP/ADP exchanger) | Required for basal and stimulated fluid secretion rates | [73] |
| Cyclic GMP-dependent kinase dg1 and dg2 (transgenic gain of function lines in tubule principal cells) | Required for stimulated fluid secretion rates | [74] |
| Dg2 for$^S$ mutant allele | Required for basal and stimulated fluid secretion rates | [75] |
| Cyclic GMP-dependent kinase dg1 (transgenic loss of function in stellate cells) | Required for diuretic-stimulated fluid secretion rates | [76] |
| Cyclic AMP-dependent kinase in principal cells<br>Cyclic AMP-dependent kinase in stellate cells | Required for basal fluid secretion rates<br>Required for stimulated fluid secretion rates | [11] |
| Exchange protein directly activated by cAMP (EPAC) | Required for stimulated fluid secretion rates | [11] |
| *Drosophila* nitric oxide synthase | Impacts on basal fluid secretion rates | [77] |
| norpA, encoding phospholipase C beta (PLC beta), and itpr, encoding inositol (1,4,5)-trisphosphate receptor (IP(3)R) | Required for neuropeptide-stimulated fluid secretion rates | [78] |
| TRPL calcium channel | Required for neuropeptide-stimulated fluid secretion rates | [79] |
| Capa receptor | Required for neuropeptide-stimulated fluid secretion rates | [80] |

(continued)

**Table 4**
**(continued)**

| Gene | Function | References |
|---|---|---|
| Kinin receptor (transgenic loss of function in stellate cells) | Required for neuropeptide-stimulated fluid secretion rates | [81] |
| Genes without effect on fluid secretion | | |
| Kinin neuropeptide (targeted knockdown in stellate cells and in abdominal neurosecretory cells (ABLKs) | No effect on neuropeptide-stimulated fluid secretion rates | [81] [82] |
| Diuretic hormone DH44 neuropeptide | No effect on neuropeptide-stimulated fluid secretion rates | [81] |
| Tyrosine decarboxylase (TDC) | No effect on fluid secretion rates | [83] |

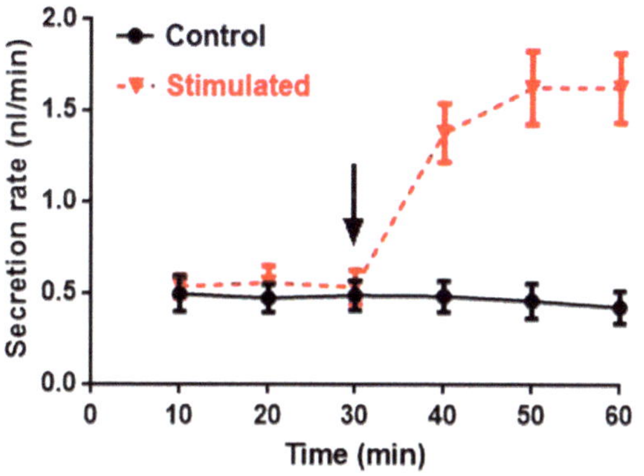


**Fig. 6** Data from a typical fluid secretion experiment from $N = 6$ tubules per condition (control and stimulated after basal rates are obtained at 30 min, arrowed). Data are shown at 10 min intervals up to 60 min and expressed as mean secretion rate (nL/min), ± SEM

## 5 Conclusion

The technique for measurement of fluid secretion rates by isolated, intact *D. melanogaster* tubules [32] described here was devised to allow physiological analysis of epithelial function in *D. Melanogaster* for the first time–thus developing integrative physiology and functional genomics in what had previously been a strictly genetic and developmental model organism. This has since allowed significant progress toward a complete physiological and molecular dissection of the Malpighian tubule, which may not be possible in any other transporting epithelium. In turn, this has allowed the development of the tubule as an important model system across multiple disciplines, including the fundamental science of understanding cellular mechanisms in epithelial function and development of new kidney disease models, and toward novel strategies for insect pest management.

## Acknowledgments

Extensive support by the Biotechnology and Biological Sciences Research Council UK to the corresponding authors has been instrumental in developing the new fields of *D. melanogaster* integrative physiology and functional genomics.

### References


1. Morgan TH (1910) Sex limited inheritance in Drosophila. Science 32:120–122
2. Rubin GM, Lewis EB (2000) A brief history of Drosophila's contributions to genome research. Science 287:2216–2218
3. Rubin GM, Spradling AC (1983) Vectors for P element-mediated gene transfer in Drosophila. Nucleic Acids Res 11:6341–6351
4. Brand AH, Perrimon N (1993) Targeted gene expression as a means of altering cell fates and generating dominant phenotypes. Development 118:401–415
5. Rosay P et al (1997) Cell-type specific calcium signalling in a Drosophila epithelium. J Cell Sci 110(Pt 15):1683–1692
6. Davies SA, Terhzaz S (2009) Organellar calcium signalling mechanisms in Drosophila epithelial function. J Exp Biol 212:387–400
7. Tian L et al (2009) Imaging neural activity in worms, flies and mice with improved GCaMP calcium indicators. Nat Methods 6:875–881
8. Shafer OT et al (2008) Widespread receptivity to neuropeptide PDF throughout the neuronal circadian clock network of Drosophila revealed by real-time cyclic AMP imaging. Neuron 58:223–237
9. Cabrero P et al (2014) Chloride channels in stellate cells are essential for uniquely high secretion rates in neuropeptide-stimulated Drosophila diuresis. Proc Natl Acad Sci U S A 111:14301-14306
10. Akerboom J et al (2013) Genetically encoded calcium indicators for multi-color neural activity imaging and combination with optogenetics. Front Mol Neurosci 6:2
11. Efetova M et al (2013) Separate roles of PKA and EPAC in renal function unraveled by the optogenetic control of cAMP levels in vivo. J Cell Sci 126:778–788
12. Dietzl G et al (2007) A genome-wide transgenic RNAi library for conditional gene inactivation in Drosophila. Nature 448:151–156
13. Gramates L et al (2017) FlyBase at 25: looking to the future. Nucleic Acids Res 45: D663–D671
14. Chintapalli VR, Wang J, Dow JA (2007) Using FlyAtlas to identify better *Drosophila melanogaster* models of human disease. Nat Genet 39:715–720
15. Leader DP, Krause SA, Pandit A, Davies SA, Dow JAT (2018) FlyAtlas 2: a new version of the Drosophila melanogaster expression atlas with RNA-Seq, miRNA-Seq and sex-specific data. Nucleic Acids Res 46:D809–D815
16. Dow JT, Davies SA (2003) Integrative physiology and functional genomics of epithelial function in a genetic model organism. Physiol Rev 83:687–729
17. Dow JAT (2012) Drosophila as an experimental organism for functional genomics. In: eLS. John Wiley & Sons Ltd, Chichester
18. Ugur B, Chen K, Bellen HJ (2016) Drosophila tools and assays for the study of human diseases. Dis Model Mech 9:235–244
19. Dow JA, Romero MF (2010) Drosophila provides rapid modeling of renal development, function, and disease. Am J Physiol Renal Physiol 299:F1237–F1244
20. Strange K (2016) Drug discovery in fish, flies, and worms. ILAR J 57:133–143
21. Sonoshita M, Cagan RL (2017) Modeling human cancers in Drosophila. Curr Top Dev Biol 121:287–309
22. Krench M, Littleton JT (2017) Neurotoxicity pathways in Drosophila models of the polyglutamine disorders. Curr Top Dev Biol 121:201–223

23. Berridge MJ, Oschman JL (1969) A structural basis for fluid secretion by malpighian tubules. Tissue Cell 1:247–272
24. Maddrell SHP (1971) The mechanisms of insect excretory systems. Adv Insect Physiol 8:199–331
25. Dow JAT (2013) In: Chapman RF, Simpson SJ, Douglas AE (eds) The insects, structure and function. Cambridge University Press, Cambridge
26. Marcelli Malpighii Philosophii & Medici Bononiensis Dissertatio epistolica De Bombyce: Societati Regiae, Londini ad Scientiam Naturalem promovendam institutae, dicata Malpighi, Marcello, 1628–1694. https://encore.lib.gla.ac.uk/iii/encore/record/C_Rb2671014;jsessionid=DD1890B70359EE4BC74ADB74D5041C5B?lang=eng
27. Maddrell S (2009) Insect homeostasis: past and future. J Exp Biol 212:446–451
28. Wessing A, Eichelberg D (1978) The genetics and biology of Drosophila, vol 2c. Academic Press, London
29. Denholm B (2013) Shaping up for action: the path to physiological maturation in the renal tubules of Drosophila. Organogenesis 9:40–54
30. Beyenbach KW, Skaer H, Dow JA (2010) The developmental, molecular, and transport biology of Malpighian tubules. Annu Rev Entomol 55:351–374
31. Sozen MA, Armstrong JD, Yang M, Kaiser K, Dow JA (1997) Functional domains are specified to single-cell resolution in a Drosophila epithelium. Proc Natl Acad Sci U S A 94:5207–5212
32. Dow JA et al (1994) The Malpighian tubules of Drosophila melanogaster: a novel phenotype for studies of fluid secretion and its control. J Exp Biol 197:421–428
33. Dube K, McDonald DG, O'Donnell MJ (2000) Calcium transport by isolated anterior and posterior Malpighian tubules of Drosophila melanogaster: roles of sequestration and secretion. J Insect Physiol 46:1449–1460
34. Dube KA, McDonald DG, O'Donnell MJ (2000) Calcium homeostasis in larval and adult Drosophila melanogaster. Arch Insect Biochem Physiol 44:27–39
35. O'Donnell MJ, Maddrell SH (1995) Fluid reabsorption and ion transport by the lower Malpighian tubules of adult female Drosophila. J Exp Biol 198:1647–1653
36. Dow JA (2009) Insights into the Malpighian tubule from functional genomics. J Exp Biol 212:435–445
37. Dow JA (1999) The multifunctional Drosophila melanogaster V-ATPase is encoded by a multigene family. J Bioenerg Biomembr 31:75–83
38. Allan AK, Du J, Davies SA, Dow JAT (2005) Genome-wide survey of V-ATPase genes in *Drosophila* reveals a conserved renal phenotype for lethal alleles. Physiol Genomics 22:128–138
39. Torrie LS et al (2004) Resolution of the insect ouabain paradox. Proc Natl Acad Sci U S A 101:13689–13693
40. Kaufmann N et al (2005) Developmental expression and biophysical characterization of a Drosophila melanogaster aquaporin. Am J Physiol Cell Physiol 289:C397–C407
41. Kerr M, Davies SA, Dow JA (2004) Cell-specific manipulation of second messengers; a toolbox for integrative physiology in Drosophila. Curr Biol 14:1468–1474
42. Wang J et al (2004) Function-informed transcriptome analysis of Drosophila renal tubule. Genome Biol 5:R69
43. Chintapalli VR et al (2012) Functional correlates of positional and gender-specific renal asymmetry in Drosophila. PLoS One 7:e32577
44. Davies SA et al (2012) Immune and stress response 'cross-talk' in the *Drosophila* Malpighian tubule. J Insect Physiol 58:488–497
45. Davies SA et al (2014) Cell signalling mechanisms in stress tolerance. J Exp Biol 217:119–128
46. Terhzaz S et al (2010) Cell-specific inositol 1,4,5 trisphosphate 3-kinase mediates epithelial cell apoptosis in response to oxidative stress in Drosophila. Cell Signal 22:737–748
47. Piermarini PM, Esquivel CJ, Denton S (2017) Malpighian tubules as novel targets for mosquito control. Int J Environ Res Public Health 14. https://doi.org/10.3390/ijerph14020111
48. Dow JA, Davies SA (2006) The Malpighian tubule: rapid insights from post-genomic biology. J Insect Physiol 52:365–378
49. Ianowski JP, O'Donnell MJ (2004) Basolateral ion transport mechanisms during fluid secretion by Drosophila Malpighian tubules: Na+ recycling, Na+:K+:2Cl− cotransport and Cl− conductance. J Exp Biol 207:2599–2609
50. Linton SM, O'Donnell MJ (1999) Contributions of K+:Cl− cotransport and Na+/K+-ATPase to basolateral ion transport in malpighian tubules of Drosophila melanogaster. J Exp Biol 202:1561–1570
51. Maddrell SH, Overton JA (1988) Stimulation of sodium transport and fluid secretion by ouabain in an insect malpighian tubule. J Exp Biol 137:265–276

52. Davies SA et al (1996) Analysis and inactivation of vha55, the gene encoding the vacuolar ATPase B-subunit in Drosophila melanogaster reveals a larval lethal phenotype. J Biol Chem 271:30677–30684
53. Karet FE et al (1999) Mutations in the gene encoding B1 subunit of H+-ATPase cause renal tubular acidosis with sensorineural deafness. Nat Genet 21:84–90
54. Kamleh MA, Hobani Y, Dow JA, Zheng L, Watson DG (2009) Towards a platform for the metabonomic profiling of different strains of Drosophila melanogaster using liquid chromatography-Fourier transform mass spectrometry. FEBS J 276:6798–6809
55. O'Donnell MJ (2009) Too much of a good thing: how insects cope with excess ions or toxins in the diet. J Exp Biol 212:363–372
56. Dow JA (2007) Integrative physiology, functional genomics and the phenotype gap: a guide for comparative physiologists. J Exp Biol 210:1632–1640
57. Tardif G, Murnik M (1975) Frequency-dependent sexual selection among wild-type strains of Drosophila melanogaster. Behav Genet 5:373–379
58. Ramsay J (1954) Active transport of water by the Malpighian tubules of the stick insect, Dixippus morosus (Orthoptera, Phasmidae). J Exp Biol 31:104–113
59. Chintapalli VR, Wang J, Herzyk P, Davies SA, Dow JA (2013) Data-mining the FlyAtlas online resource to identify core functional motifs across transporting epithelia. BMC Genomics 14:518
60. Du J et al (2006) The SzA mutations of the B subunit of the Drosophila vacuolar H+ ATPase identify conserved residues essential for function in fly and yeast. J Cell Sci 119:2542–2551
61. Maddrell SH, O'Donnell MJ (1992) Insect Malpighian tubules: v-ATPase action in ion and fluid transport. J Exp Biol 172:417–429
62. Coast GM, Webster SG, Schegg KM, Tobe SS, Schooley DA (2001) The *Drosophila melanogaster* homologue of an insect calcitonin-like diuretic peptide stimulates V-ATPase activity in fruit fly Malpighian tubules. J Exp Biol 204:1795–1804
63. Giannakou ME, Dow JA (2001) Characterization of the Drosophila melanogaster alkali-metal/proton exchanger (NHE) gene family. J Exp Biol 204:3703–3716
64. Davies SA, Day JP (2006) cGMP signalling in a transporting epithelium. Biochem Soc Trans 34:512–514
65. Coast G (2007) The endocrine control of salt balance in insects. Gen Comp Endocrinol 152:332–338
66. Blumenthal EM (2003) Regulation of chloride permeability by endogenously produced tyramine in the Drosophila Malpighian tubule. Am J Physiol Cell Physiol 284:C718–C728
67. Wu Y, Schellinger JN, Huang CL, Rodan AR (2014) Hypotonicity stimulates potassium flux through the WNK-SPAK/OSR1 kinase cascade and the Ncc69 sodium-potassium-2-chloride cotransporter in the Drosophila renal tubule. J Biol Chem 289:26131–26142
68. O'Donnell MJ, Ianowski JP, Linton SM, Rheault MR (2003) Inorganic and organic anion transport by insect renal epithelia. Biochim Biophys Acta 1618:194–206
69. Halberg KA et al (2016) The cell adhesion molecule Fasciclin2 regulates brush border length and organization in Drosophila renal tubules. Nat Commun 7:11266
70. Wu Y, Baum M, Huang CL, Rodan AR (2015) Two inwardly rectifying potassium channels, Irk1 and Irk2, play redundant roles in Drosophila renal tubule function. Am J Physiol Regul Integr Comp Physiol 309:R747–R756
71. Rodan AR, Baum M, Huang CL (2012) The Drosophila NKCC Ncc69 is required for normal renal tubule function. Am J Physiol Cell Physiol 303:C883–C894
72. Day JP et al (2008) Identification of two partners from the bacterial kef exchanger family for the apical plasma membrane V-ATPase of Metazoa. J Cell Sci 121:2612–2619
73. Terhzaz S, Cabrero P, Chintapalli VR, Davies SA, Dow JAT (2010) Mislocalization of mitochondria and compromised renal function and oxidative stress resistance in Drosophila SesB mutants. Physiol Genomics 41:33–41
74. MacPherson MR, Lohmann SM, Davies SA (2004) Analysis of Drosophila cGMP-dependent protein kinases and assessment of their in vivo roles by targeted expression in a renal transporting epithelium. J Biol Chem 279:40026–40034
75. MacPherson MR et al (2004) The dg2 (for) gene confers a renal phenotype in Drosophila by modulation of cGMP-specific phosphodiesterase. J Exp Biol 207:2769–2776
76. Ruka KA, Miller AP, Blumenthal EM (2013) Inhibition of diuretic stimulation of an insect secretory epithelium by a cGMP-dependent protein kinase. Am J Physiol Renal Physiol 304:F1210–F1216
77. Broderick KE et al (2003) Interactions between epithelial nitric oxide signaling and

phosphodiesterase activity in Drosophila. Am J Physiol Cell Physiol 285:C1207–C1218

78. Pollock VP et al (2003) NorpA and itpr mutants reveal roles for phospholipase C and inositol (1,4,5)- trisphosphate receptor in Drosophila melanogaster renal function. J Exp Biol 206:901–911
79. MacPherson MR et al (2005) Transient receptor potential-like channels are essential for calcium signaling and fluid transport in a Drosophila epithelium. Genetics 169:1541–1552
80. Terhzaz S et al (2012) Mechanism and function of *Drosophila* capa GPCR: a desiccation stress-responsive receptor with functional homology to human neuromedinU receptor. PLoS One 7:e29897
81. Cannell E et al (2016) The corticotropin-releasing factor-like diuretic hormone 44 (DH) and kinin neuropeptides modulate desiccation and starvation tolerance in Drosophila melanogaster. Peptides 80:96–107
82. Zandawala M, Marley R, Davies SA, Nassel DR (2018) Characterization of a set of abdominal neuroendocrine cells that regulate stress physiology using colocalized diuretic peptides in Drosophila. Cell Mol Life Sci 75:1099–1115
83. Blumenthal EM (2009) Isoform- and cell-specific function of tyrosine decarboxylase in the Drosophila Malpighian tubule. J Exp Biol 212:3802–3809
84. Ashburner M (1989) Drosophila: A laboratory handbook. Cold Spring Harbor Laboratory: 1331
85. Maddrell SHP (1991) BioEssays 13(7):357

# Chapter 18

# Analysis of the Teashirt Target Genes in Ureteric Bud Development

**Laurent Fasano, Irene Sanchez-Martin, and Xavier Caubit**

## Abstract

Microarrays and RNA-seq (RNA sequencing) are powerful techniques to assess transcript abundance in biological samples and to improve our understanding of the relationship between genotype and phenotype. *Tshz3*$^{+/-}$ heterozygous mouse is a model for a human 19q12 syndrome characterized by autistic traits and renal tract defects (Caubit et al., Nat Genet 48:1359–1369, 2016). To unravel renal tract pathological mechanisms, we took advantage of *Tshz3* mouse and performed comparative genome-wide expression profiling on embryonic ureter and/or kidney.

**Key words** Microarray, RNA-seq, Mouse, Ureter, Kidney, *Tshz3*

## 1 Introduction

During the last three decades, the rapid development of genomic tools has offered the possibility to perform genome-wide analyses to identify changes associated with gene mutation in human and model organisms. In particular, transcriptomic analyses using microarrays and RNA-seq have improved our understanding of the molecular and cellular mechanisms occurring during the normal and pathological development of different model systems.

Historically, miniaturized microarrays prepared by high-speed robotic printing of complementary DNAs on glass began in the mid-1990s [1], and the first complete eukaryotic genome was placed on DNA chip in 1997 [2]. Since its first research use, different innovations (i.e., fabrication, robotics, informatics, and bioinformatics) have dramatically improved the efficiency of microarray and reduced the cost of this technology. Nowadays and despite the improvements mentioned above, microarrays are challenged by RNA-seq, a high-throughput DNA sequencing method, allowing mapping and quantifying transcriptomes [3–5]. Indeed, RNA-seq, which uses next-generation sequencing (NGS) technologies, presents several advantages compared to

Seppo Vainio (ed.), *Kidney Organogenesis: Methods and Protocols*, Methods in Molecular Biology, vol. 1926, https://doi.org/10.1007/978-1-4939-9021-4_18, 

microarray. It does not depend on existing knowledge about genome sequence (in other words "you don't need to know what you sequence"), it has very low background, it allows quantification of expression levels, it is highly accurate, and the results are very reproducible.

Both microarrays and RNA-seq have revolutionized the way we probe whole-genome transcriptome, enabling us to generate, analyze, and explore the transcriptome for a number of species, organs, and cell types. In developmental biology, these technologies are methods of choice to profile differentially expressed genes between different experiment conditions (i.e., healthy vs. mutant tissues; time point 1 vs. time point 2), which allow understanding transcriptomic dynamics in time and space and comparison between mutant and normal conditions.

We combined mouse genetics and gene expression profiling to provide new insight into the role of differential gene expression in normal and pathological renal tract development. More specifically, we took advantage of *Teashirt 3* (*Tshz3*) mutant mice that exhibit congenital pelviureteric junction obstruction (PUJO) with defective smooth muscle differentiation and absent peristalsis in the proximal ureter [6, 7]. PUJO affects 0.3% of human births, and in human, *TSHZ3* is expressed in peri-urothelial cells of the proximal ureter and renal pelvis [8]. More recently, *TSHZ3* was identified as a gene candidate for a syndrome characterized by autistic traits and renal tract defects [9]. Therefore, deciphering the molecular functions of *Tshz3* would be valuable to understand the basis of this syndrome. To this aim, we combined mouse genetics and gene expression profiling to provide new insight into the role of differential gene expression in normal and pathological renal tract development. We used microarray to evaluate up-regulated or down-regulated genes in *Tshz3*$^{lacZ/lacZ}$ mutant uretersureters. The analysis was performed at mouse Theiler stage 22 (TS22; Embryonic stage (E) 14.5), which corresponds to the onset of the myogenic program and, at TS24-25 (E16.5), when smooth muscle cells (SMCs) express the full repertoire of differentiation marker genes. Recently, we have examined the differences in mRNA expression in TS20 (E12.5) *Tshz3*$^{lacZ/lacZ}$ mutant ureters and/or kidneys using RNA-seq.

## 2 Materials

### 2.1 Dissection and Sample Collection

1. Dissecting microscope with transillumination from below and fiber optic from above.
2. Petri dishes (35 mm or 65 mm) filled with silicone.
3. Disposable plastic container with disposable stir stick.
4. Tungsten needles inserted into a needle holder.

5. Dissecting instruments: scissors, microdissecting scissors, and forceps.
6. Phosphate-buffered saline (PBS; 1×). To prepare 1 L of 1× PBS, dissolve 8 g of NaCl, 0.2 g of KCl, 1.44 g of $Na_2HPO_4$, and 0.24 g of $KH_2PO_4$ in 800 mL of $H_2O$. Adjust the pH to 7.4 with HCl. Add $H_2O$ to 1 L (*see* **Note 1**).
7. Liquid nitrogen or dry ice (*see* **Note 2**).

### 2.2 mRNA Isolation

1. Lab protocol for mRNA isolation or kit/reagents from your favorite manufacturer.
2. Ad hoc machines to control RNA quality, integrity, and purity (*see* **Note 3**).
3. 26G needles.
4. Eppendorf tubes.
5. RNase-, DNase-, and protease-free molecular grade water (*see* **Note 3**).

## 3 Methods

### 3.1 Dissection and Sample Collection

1. *Tshz3* mutant and wild-type kidneys and ureters were collected at E12.5 for RNA-seq and at E14.5 and E16.5 for microarrays.
2. For the dissection, embryos were placed on petri dishes (35 mm or 65 mm) containing a silicone pad and covered with 1× PBS (*see* **Note 4**).
3. Dissections were performed mainly as described in "Dissection of Embryonic Mouse Kidney, Culture In Vitro, and Imaging of the Developing Organ" [10] and "Mouse Urogenital Development: A Practical Approach" [11]. For E12.5 samples, we adapted the protocol to improve the speed of the dissection (*see* **Note 5**).
4. E12.5 samples were placed in an Eppendorf directly in liquid nitrogen (*see* **Note 2**) and kept frozen at −80 °C until RNA extraction. Upon dissection, E14.5 and E16.5 ureters were immediately dropped into ready-to-use reagent for the isolation of total RNA according to manufacturer's instructions.

### 3.2 mRNA Isolation

1. Tissue was homogenized using a 26-gauge needle attached to a sterile plastic syringe.
2. Total RNA was extracted using a RNA extraction kit according to manufacturer's instructions. The procedure allows (1) to lyse the tissue with a denaturing guanidine-isothiocyanate-containing buffer, which inactivates RNAses to prevent RNA degradation, and (2) to eliminate the DNA, ribosomal RNA, and tRNA (*see* **Note 6**).

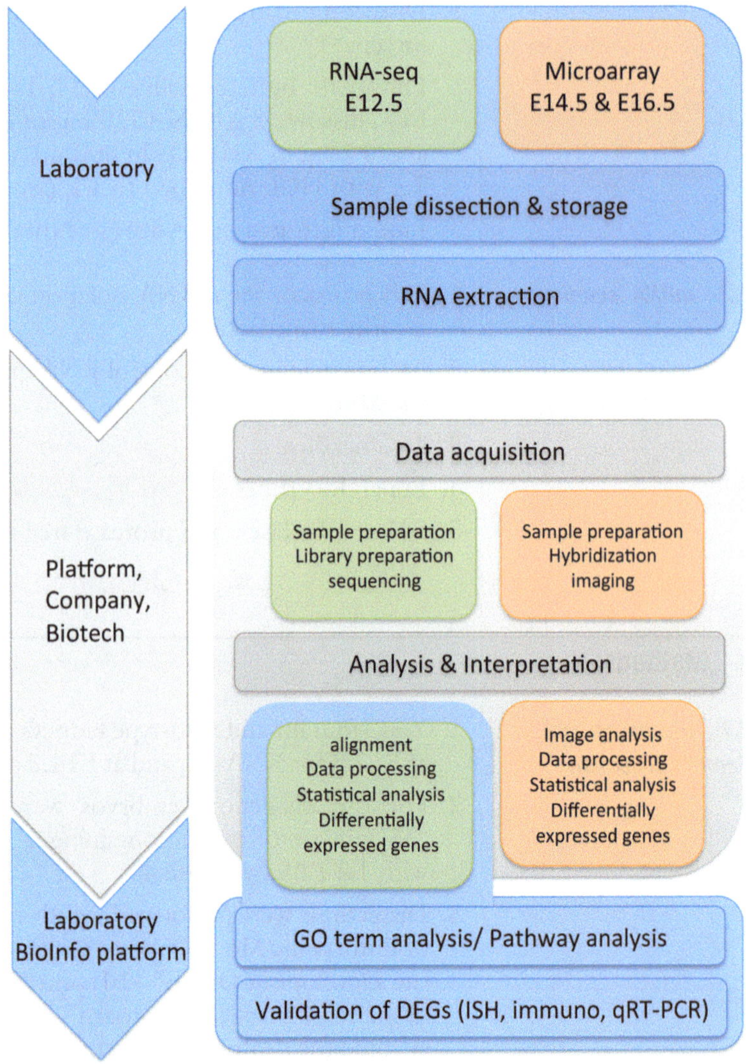


**Fig. 1** Workflow of samples for RNA-seq (left) and microarray (right)

3. The integrity of RNA was assessed using a chip-based capillary electrophoresis machine and RNA concentration determined using a full-spectrum (220–750 nm) spectrophotometer (*see* **Note 3**) (Fig. 1).

### 3.3 Data Acquisition Microarrays

1. *Label protocol*: cRNAs for hybridization to "Provider A" arrays were prepared from 100 ng total RNA using the wild-type expression kit from "Provider A."
2. *Hybridization protocol*: Labelled cRNA was fragmented and hybridized to mouse gene arrays (Provider A) following manufacturer's protocols. Three independent RNA preparations (triplicate) from the four different conditions [wild type and

mutant at E14.5; wild type and mutant at E16.5] were processed and hybridized on mouse gene arrays.

3. *Data processing*: Arrays have been quantile-normalized with respect to the probe GC content using the RMA algorithm (GC contend adjustment, RMA background correction, and mean probe set summarization). Not or low expressed transcripts have been removed by a maximum expression cutoff <100. The data filtering results in 173,540 of 234,872 probe sets and 22,495 meta-probe sets. After normalization the arrays have been checked for outlier using the principal component analysis using a correlation dispersion matrix and normalized eigenvector scaling. No outlier has been detected. Microarray probe sets representing transcripts changed by twofold or more were further analyzed. Differential expression of summarized gene level expression was calculated using the Partek two-way ANOVA statistic followed by a FDR multiple testing correction. Resulting p-values and FDR values indicate the probability of differential expression in state (wild type/knockout) and stage (embryonic stages 14 and 16). Two thousand two hundred and forty transcript clusters with an ANOVA FDR < 0.05 in one of the conditions (state, stage, state-stage interaction) have been defined as the significant differential expressed subset. Data have been deposited in the Gene Expression Omnibus (GEO) under accession GSE43940. Series record GSE43940 provides access to all of our data and is the accession that should be quoted in any manuscript discussing the data.

### 3.4 Analysis and Interpretation

We have four conditions that we can compare: E14.5 WT, E16.5 WT, E14.5 *Tshz3*$^{lacZ/lacZ}$ (E14.5 KO), and E16.5 *Tshz3*$^{lacZ/lacZ}$ (E16.5 KO). We previously reported that myocardin (*Myocd*), a master regulator of the smooth muscle gene expression [12], was not expressed in the *Tshz3*$^{lacZ/lacZ}$ mutant ureteral mesenchyme [6]. We used *Myocd* as an internal control (*see* **Note 7**). In both E14.5 *Tshz3*$^{lacZ/lacZ}$ compared to E14.5 WT (E14.5 KO vs. WT) and E16.5 *Tshz3*$^{lacZ/lacZ}$ compared to E16.5 WT (E16.5 KO vs. WT) ureters, *Myocd* showed the third strongest downregulation; thus for these two conditions, statistically significant ($p < 0.05$) and differentially expressed genes (DEGs) were considered for future investigation (*see* **Note 7**). Among the 22,495 genes present on the mouse GeneChip, we identified 2311 DEGs in WT E16.5 vs. E14.5 and 3614 genes in KO E16.5 vs. E14.5. Using an absolute fold change cutoff value of 1 in $\log_2$ scale, we identified 1112 downregulated and 1199 upregulated genes in WT E16.5 vs. E14.5 and 1851 downregulated and 1763 upregulated genes in KO E16.5 vs. E14.5. To characterize the dynamic regulation of gene expression, we used Venny2.1 [13]. This approach allowed comparing the gene expression patterns and showed that the absence of *Tshz3* leads to a massive modification of gene

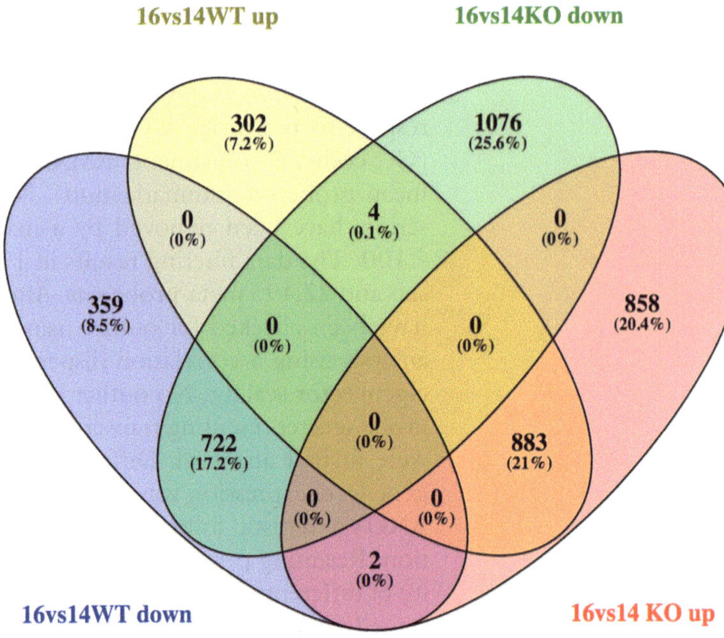


**Fig. 2** Differentially expressed genes in wild-type and mutant ureters. Statistically significant differentially expressed genes in the comparison made between WT E16.5 vs. E14.5 and KO E16.5 vs. E14.5

regulation. The comparative analysis of the two conditions identified 4206 DEGs: 2049 being upregulated and 2163 downregulated. 15.7% of the DEGs were unique to WT E16.5 vs. E14.5 (302 upregulated and 359 downregulated), and 46% were exclusively found in KO E16.5 vs. E14.5 (858 upregulated and 1076 downregulated). 38.2% were common to both conditions (883 upregulated and 722 downregulated). Lastly, only six genes showed inverted patterns: four were up in WT and down in KO, and two were down in WT and up in KO (*see* **Note 8**) (Fig. 2).

### 3.5 Data Acquisition RNA-Seq

1. *Random-primed cDNA library*: Starting material, 1 μg Total mRNA (minimum 500 ng; concentration > 20 ng/μL); purity, OD 260/280, 1.8; RNA integrity number (RIN) ≥ 8 (*see* **Note 3**). Dissolved in RNase-, DNase-, and protease-free molecular grade water (*see* **Note 9**). To obtain two independent total RNA samples from the two different conditions (wild type, WT1 and WT2; mutant, KO1 and KO2), we dissected and pooled a minimum of eight ureters in each preparation. Upon purification, poly-A-containing mRNA was fragmented, and cDNA was prepared by random-primed synthesis.
2. *Sequencing:* Run type, pair end; read length, 2 × 50 bp.

Sample (S) WT1, sequenced reads (SR) 38,564,100; S WT2, SR 52,609,430; S KO1, SR 65,805,700; and S KO2, SR 49,445,500.

3. *Data processing*: The raw data (sequences; fastq format) (*see* **Note 10**) were mapped to the mm10 version of the mouse genome to generate Sequence Alignment/Map (SAM/BAM) format. SAM is a tab-delimited alignment format consisting of a header section and an alignment section with 12 columns. BAM is the compressed, indexed, and binary version of SAM format. SAM/BAM are also essential and precious, because this is what is mapped and hence used to count reads. After normalization, analysis of differentially expressed genes (DEGs) was performed using both the Bioconductor package DESeq/DESeq2 and the package edgeR. This analysis generated differential expression lists with false discovery rates (FDRs), which are derived from p-values corrected for multiple testing using the Benjamini-Hochberg method. Six files in total were generated: FDR 1%, 5%, and 10% for both UCSC (transcripts) and ENSEMBL (genes).

### 3.6 Analysis and Interpretation of RNA-Seq Data

These data are currently under investigation to validate potential differentially expressed genes (DEGs) using in situ hybridization (ISH), immunodetection, and/or qRT-PCR.

## 4 Notes

1. Dispense the solution into aliquots, and sterilize them by autoclaving for 20 min at 15 psi (1.05 kg/cm$^2$) on liquid cycle or by filter sterilization. Store at room temperature. You can also follow the same protocol to prepare a 10× PBS stock solution using 80 g of NaCl, 2 g of KCl, 14.4 g of $Na_2HPO_4$, and 2.4 g of $KH_2PO_4$.
2. Because of its properties (extremely cold boiling point −196 °C), small volumes vaporize to give large volumes of gas (1 L gives 0.7 m$^3$ of gas) which will displace oxygen in air; liquid nitrogen can cause sever burn-like damage to the skin, and the low temperature of the vapor can cause damage to softer tissues (e.g., eyes, lungs); it is highly recommended to know how to transport and handle liquid nitrogen.
3. To determine RNA quantity and purity, the best way is to measure UV absorption of sample using a spectrophotometer requiring 1–2 μL of sample. Ideally, this spectrophotometer contains preset program RNA measurements that automatically read the OD at 260, 280, and 230. The concentration of the RNA can be determined by measuring the $OD_{260}$. A solution of RNA whose $OD_{260} = 1$ contains ±40 μg of RNA

per milliliter. The $OD_{260}/OD_{280}$ ratio is an indication of the level of protein contamination in your sample. Pure RNA has an $OD_{260}/OD_{280}$ ratio of 2.1 ($1.8 < OD_{260}/OD_{280} < 2.0$ are considered acceptable). Depending on the isolation protocol you used to extract RNA, your sample can also contain contaminant (e.g., guanidine salts, phenol). A high peak at $OD_{230}$ indicates contamination. $1.5 < OD_{260}/OD_{230}$ is considered as an ideal ratio.

Remember the best measurements are obtained (1) when RNA is solubilized in Tris EDTA (TE) buffer, (2) the blank solution is the same solution the RNA is diluted in, and (3) RNA concentration below 20 μg/mL does not give reliable measurements.

To check for RNA integrity, the best is to use a Bioanalyzer using a small amount of RNA (1–2 μL). Moreover, the analyzer determines an RNA integrity number (RIN) to standardize between RNA samples. These apparatus are expensive. If you don't have access to an analyzer, you can still run the RNA on a 1% standard agarose gel and examine the ribosomal RNA (rRNA). mRNA runs between the two ribosomal bands (28S–18S in eukaryotic cells; 23S–16S in bacteria cells) and might be seen as a smear. A smear below the lower rRNA band (18S or 16S) suggests that the RNA is of poor quality. If you find that your RNA is degraded, you should prepare new RNA for your experiment.

4. To prepare the pad, use a disposable plastic container and a stir stick to make the elastomer (silicone) according to the manufacturer's instructions. Pour the elastomer into petri dishes, filling them a third to half way through. Let dry the dishes covered for ±24 h at room temperature.
5. For dissecting the E12.5 kidney/ureter, we cut embryos below the upper arms to keep the posterior half. We eliminated all the structures except the gonads and renal tract that are attached to the back. Kidneys and ureters were dissected using forceps and/or needles and put in an Eppendorf tube using a small spoon.
6. RNA purification should follow tissue isolation as quickly as possible. If not possible immerse the tissues in a RNA stabilization solution, or freeze them (i.e., liquid nitrogen, dry ice) to prevent alteration in the transcript profile. Samples for a given experiment should be collected following the same protocol.
7. We noticed that *Myocd* levels were not significantly affected in E16.5 WT compared to E14.5 WT (WT E16.5 vs. E14.5) ureters, suggesting that at E14.5 *Myocd* expression has already reached a plateau. This result was also observed in E16.5 *Tshz3*$^{lacZ/lacZ}$ compared to E14.5 *Tshz3*$^{lacZ/lacZ}$ (KO E16.5 vs. E14.5) ureters. For these two conditions, in

order to keep *Myocd* in our analysis, we had to use a $p < 0.057$ in E14.5 KO vs. WT and $p < 0.055$ in E16.5 KO vs. WT, indicating that in vivo data could be helpful to choose a threshold when performing a global analysis.

8. The data mentioned here are accessible through GEO Series accession number GSE43940 at http://www.ncbi.nlm.nih.gov/geo/query/acc.cgi?acc=GSE43940. Partial analysis of the data was previously published by Denholm et al. Development 140, 1100–1110, 2013 (PMID 23404107).
9. The best is probably to buy a RNase-, DNase-, and protease-free molecular grade water from a provider. Otherwise you should use "ultrapure" Milli-Q (reverse osmosis purified) water or double distilled water. Add 1 mL DEPC (diethyl pyrocarbonate) per 1 L of water (i.e., to a final concentration 0.1%). Mix thoroughly, and let incubate for 12 h at 37 °C. Autoclave DEPC-mixed water for 15 min. DEPC inactivates RNases by covalent modification, and 0.1% DEPC is probably sufficient to inactivate most RNase contamination (the efficacy of the DEPC in inactivating the RNase depends on the concentration of RNase, but increasing the concentration of DEPC can affect the quality of RNA). The autoclaving step inactivates the DNases and removes all traces of DEPC.
10. One backup copy of the raw data (sequences) should be kept preciously and never use for analysis.

## References


1. Caubit X, Gubellini P, Andrieux J, Roubertoux PL, Metwaly M, Jacq B, Fatmi A, Had-Aissouni L, Kwan KY, Salin P, Carlier M, Lieden A, Rudd E, Shinawi M, Vincent-Delorme C, Cuisset JM, Lemaitre MP, Abderrehamane F, Duban B, Lemaitre JF, Woolf AS, Bockenhauer D, Severac D, Dubois E, Zhu Y, Sestan N, Garratt AN, Kerkerian-Le Goff L, Fasano L (2016) TSHZ3 deletion causes an autism syndrome and defects in cortical projection neurons. Nat Genet 48(11):1359–1369. https://doi.org/10.1038/ng.3681
2. Schena M, Shalon D, Davis RW, Brown PO (1995) Quantitative monitoring of gene expression patterns with a complementary DNA microarray. Science 270(5235):467–470
3. Lashkari DA, DeRisi JL, McCusker JH, Namath AF, Gentile C, Hwang SY, Brown PO, Davis RW (1997) Yeast microarrays for genome wide parallel genetic and gene expression analysis. Proc Natl Acad Sci U S A 94 (24):13057–13062
4. Mortazavi A, Williams BA, McCue K, Schaeffer L, Wold B (2008) Mapping and quantifying mammalian transcriptomes by RNA-Seq. Nat Methods 5(7):621–628. https://doi.org/10.1038/nmeth.1226
5. Nagalakshmi U, Wang Z, Waern K, Shou C, Raha D, Gerstein M, Snyder M (2008) The transcriptional landscape of the yeast genome defined by RNA sequencing. Science 320 (5881):1344–1349. https://doi.org/10.1126/science.1158441
6. Wilhelm BT, Marguerat S, Watt S, Schubert F, Wood V, Goodhead I, Penkett CJ, Rogers J, Bahler J (2008) Dynamic repertoire of a eukaryotic transcriptome surveyed at single-nucleotide resolution. Nature 453 (7199):1239–1243. https://doi.org/10.1038/nature07002
7. Caubit X, Lye CM, Martin E, Core N, Long DA, Vola C, Jenkins D, Garratt AN, Skaer H, Woolf AS, Fasano L (2008) Teashirt 3 is necessary for ureteral smooth muscle differentiation downstream of SHH and BMP4. Development

135(19):3301–3310. https://doi.org/10.1242/dev.022442

8. Martin E, Caubit X, Airik R, Vola C, Fatmi A, Kispert A, Fasano L (2013) TSHZ3 and SOX9 regulate the timing of smooth muscle cell differentiation in the ureter by reducing myocardin activity. PLoS One 8(5):e63721. https://doi.org/10.1371/journal.pone.0063721
9. Jenkins D, Caubit X, Dimovski A, Matevska N, Lye CM, Cabuk F, Gucev Z, Tasic V, Fasano L, Woolf AS (2010) Analysis of TSHZ2 and TSHZ3 genes in congenital pelvi-ureteric junction obstruction. Nephrol Dial Transplant 25 (1):54–60
10. Costantini F, Watanabe T, Lu B, Chi X, Srinivas S (2011) Dissection of embryonic mouse kidney, culture in vitro, and imaging of the developing organ. Cold Spring Harb Protoc 2011(5):pdb prot5613. https://doi.org/10.1101/pdb.prot5613
11. Staack A, Donjacour AA, Brody J, Cunha GR, Carroll P (2003) Mouse urogenital development: a practical approach. Differentiation 71 (7):402–413. https://doi.org/10.1046/j.1432-0436.2003.7107004.x
12. Wang Z, Wang DZ, Pipes GC, Olson EN (2003) Myocardin is a master regulator of smooth muscle gene expression. Proc Natl Acad Sci U S A 100(12):7129–7134. https://doi.org/10.1073/pnas.1232341100
13. Oliveros JC (2007–2015) Venny. An interactive tool for comparing lists with Venn's diagrams. http://bioinfogp.cnb.csic.es/tools/venny/

# Part VII

## Computational Approaches to Kidney Ontogenesis

# Chapter 19

# Methods of Computational Analysis in Kidney Development

**Pauli Tikka and Franz Schaefer**

## Abstract

This chapter reviews some currently available methodologies for constructing mathematical models in kidney development. Mammalian nephrogenesis is a complex biological process, which in its earliest stages involves migration, condensation, proliferation, and differentiation of metanephric mesenchymal (MM) cells interacting with the uroepithelial cells of the ureteric bud (UB). First, the mathematical modelling in biology is generally described. Secondly, some accounts to biological pattern formation in modelling are given in general, including models that transcend the Turing model. This is followed by a short assessment on the main branch of models in the kidney development, the evaluation of the branching morphogenesis of the kidney. Finally, two alternative models in the early kidney development processes are given as an example. They also elucidate the difficulties in the model building process. Firstly, a computational model building with the CompuCell3D program for the early nephron progenitor cell movements with the key extracellular signaling effectors is depicted. This collective migration leads to the first pretubular aggregate (PTA). The simulation parameters of the program imitate the program's cell sorting example with different adhesions and chemoattractants. The program utilizes Cellular Potts Model (CPM) to describe the development. Secondly, an example of PTA to renal vesicle (RV) transition modelling is described. In that case, the model is unique, where the model process is based on the chemoattractants from UB.

**Key words** Modelling in biology, Computational modelling, CompuCell3D, Early nephrogenesis

## 1 Introduction

In order to define an observed feature of the nature more clearly, scientific modelling aims to reface it to existing and commonly accepted knowledge and quantify it [1]. It requires selecting and identifying relevant aspects of the situation in hand and then using different types of models for different aims. In particular, mathematical models try to describe experimentally delineated phenomenon by means of mathematics, with a view to capturing the salient aspects of the phenomenon at hand [2]. They consist of relationships that can be described by operators, such as functions, and variables that are abstractions of the parameters of interest. Mathematical modelling approaches can be roughly divided into six

Seppo Vainio (ed.), *Kidney Organogenesis: Methods and Protocols*, Methods in Molecular Biology, vol. 1926,
https://doi.org/10.1007/978-1-4939-9021-4_19, 

categories according to the form of the relationships and variables: linear vs. nonlinear, static vs. dynamic, explicit vs. implicit, discrete vs. continuous, deterministic vs. probabilistic, and deductive vs. inductive or floating [1]. The actual mathematical model is the predictive explanation arising from the modelling. The most important tools for biological modelling are differential equations, probability, and statistics, i.e., bioinformatics. The model building work process consists roughly of five stages: (1) defining the problem, (2) creating the model, (3) developing a computational mode for solving the model, (4) implementation of the model, and (5) testing and assessing the solution. The work process can consist of multiple trial-and-error model revisions if the data fed to the model are inconsistent or the assumptions behind the model were not adequately explored in the beginning.

In the field of kidney development, the computational efforts around ureteric branching morphogenesis, interwoven to the early nephron formation in various levels, have advanced various models in vivo, ex vivo, in vitro, and in silico [3–6]. The patterns of cells have been conceived to arrive from signaling and evolving membrane interactions from the ureteric, mesenchymal, interstitial, and nephron progenitor cells [7, 8]. Lawson et al. created a mathematical model for the induction of the mammalian ureteric bud [3]. Zubkov et al. established a spatially averaged mathematical model of kidney branching morphogenesis [9]. Regardless, these models did not account for the spatiotemporal evolution of nephron progenitor (NP) cells and nephron development. Accordingly, refined computational approaches with CompuCell3D are shortly delineated here [10, 11]. In these approaches, NPs are sorted and drawn from MM cells toward PTAs via extracellular Wnt9b morphogens and cell adhesion differences [12, 13]. The process starts after the first UB has branched, and NPs are about to move, E11.5.

## 2 Mathematical Modelling in Developmental Biology

### 2.1 Modelling in Biology

A brief account in the mathematical modelling in the developmental biology and in the kidney development is described. Other types of modelling and models, such as model organisms, physical/scale models, nor the common discussion of the concept of the models, are not covered here. Mathematical modelling in biology is the describing of an experimentally delineated biological phenomenon by means of mathematics. Mathematical models can recapitulate the underlying mechanisms of the biological processes and reduce the large data sets of the experimental measurements to fewer and more meaningful ones. The parameter values of the model can be estimated with the data. As well as, a model is confronted with the data when testing whether the model predictions concord with the data by fitting the model to the data. However, the models are not just for processing data, rather than for determining the best and most relevant experiments [2].

The model building process sharpens one's understanding of the involved hypotheses. During the model building process, one should strive for a long list of simple assumptions. Assumptions can be categorized into implicit vs. explicit, strong vs. weak, and stringent vs. relaxed ones [2]. This allows a systematic exploration of variations in the model, as well as comparison among closely related version. The statistics, probability calculations, and differential equations (DEs) are the most used mathematical technique for describing the biological processes which change within time and space, such as the blood volumes in the glomerular capillaries. The questions in the developmental biology pattern formations are often evaluated with image analysis and numerical simulations, such by using Fiji and CompuCell3D programs, rather than analytically [14].

The fitting of model parameters to the experimental ones can be performed in many ways. As an example, the sum of squared errors from the true values can give an appropriate "goodness of fit" to the model. If the model involves more than one variable and several experimental data sets are available, the model can use all the available data sets at once. This is called simultaneous fitting, and it is better than fitting the individual data sets to purely descriptive convenient functions, such as straight lines. One example of simultaneous fitting is to model the carbon fixation rate in photosynthesis with two different carbon dioxide concentrations as a function of light intensity [2].

### 2.2 Modelling of Biological Patterns

Ever since the famous publication of Alan Turing, the developmental biology community has had a reference point for the modelling in biology [15]. It suggested that a system of morphogens is adequate enough to account for the main phenomena of morphogenesis. This system may later develop a pattern due to instability of the homogenous equilibrium, which is triggered off by random disturbances. Slightly a different aspect was developed later by Gierer and Meinhardt, who formed a theory based on short-range activation and long-range inhibition [16]. This theory also separates the initial morphogen concentration at the source and the effective concentration at the recipient. But the systems can be also cyclic as shown by Schnakenberg in 1978 [17]. He sought the most simple chemical reaction system which is capable of producing a limit cycle oscillations. Gierer and Meinhardt continued with evaluating structures during morphogenesis via generation and regeneration of sequences of the elements in space [18]. The general properties of these interactions were demonstrated by a system of two components which mutually activated each other. It yielded a symmetric double gradients or periodic patterns. An example of monotonic gradient was considered for the cell's response to morphogenetic fields. In that model, the regulation and regeneration of sections and transplants was a consequence of the reformation of the morphogen gradient. These structures could be also viewed as topologically distinguished maps which was later analyzed by Kohonen [19]. This approach has clarified the modelling in the self-organization of the features of the models not just related to the developmental biology.

### 2.3 Beyond the Turing Model

The mesenchymal morphogenesis, not only as in the early nephrogenesis, has been modelled with mechanical models by Murray et al. [20]. They considered a population of cells moving with an elastic medium which they associated with an initially isotropic extracellular matrix, where, if not otherwise effected, the cells demonstrate a random walk. This randomity was biased by the tendency of the cells to align along anisotropies in the substratum and move up to gradient from the substratum's adhesive sites. Murray et al. discovered that the formed equations could predict the motile cell populations spontaneously forming spatial structures by a sequence of bifurcations similar to nonlinear diffusion-reaction systems. Another landmark work was Murray's review on mechanochemical models for generating biological pattern and form in development. These models were discussed in the context of the skin organ patterns, cartilage condensations in the developing vertebrate limb, and wound healing [21]. Meinhardt continued with Koch to discuss the matters of biological pattern formation in more deep fashion [22]. In that discussion, they delineated the previously mentioned issues in activator-substrate/inhibitor systems, but also conversed biological switches, regeneration properties and polarity, and animal coat patterns, and finally the positioning mechanisms during plant growth, which is still an important open question. Based on these principles, Plaza et al. described a general model to describe the spatiotemporal dynamics of two morphogens [23]. Their diffusion process included spatiotemporal varying diffusion coefficients, advection, and dilution terms. They presented specific examples by analyzing a third-order activator-inhibitor mechanism for the kinetic part.

Nevertheless, motivated by the current experimental discoveries, Mercker et al. proposed a novel mechanism of embryonic pattern formation based on coupling of tissue curvature with diffusive signaling by a chemical factor [24]. They derived a novel model using energy minimization approach and showed that the model makes a multiple of morphogen and curvature patterns similar to observed ones. Important in this article was that the mechanism was different than the classical Turing one. Instead, their studies showed that the mechanical forces replace the long-range inhibitor and lead to formation of heterogeneous structures without existence of chemical prepatterns. Later on they also numerically investigated mechanochemical processes, in which morphogen dynamics and tissue mechanics are coupled by feedback loops [25, 26]. And as earlier found, simple mechanochemical interaction terms were sufficient to create de novo patterns.

### 2.4 Kidney Branching Morphogenesis

The most extensively modelled kidney development process has been the kidney branching morphogenesis [3, 9, 27–29]. Starting from the interaction processes between branching UB and surrounding MM cell populations, which were studied with a spatially averaged mathematical model [9]. The cell growth and maturation rates were assumed to be governed in the model by two ordinary differential equations (ODE). The model was developed to estimate the number of branch generations. It was found that MM cells increase branch number more than UB cells and indicated that the cessation of branching requires an active trigger. The model also predicted the branching outcome when the cells changed their behavior in response to different genetic or environmental developmental stresses. Nevertheless, this approach could not solve the relationship between the tip and the final nephron number and also why some MM cells are randomly swarming between UB cells.

The image analysis-based modelling of branching usually starts by recording it to 2D video as in the works of Adivarahan et al. [27]. It is followed by the calculation of the cell cluster displacements fields of the consecutive images of the videos with MATLAB or other mathematical programs [30]. These fields can be then imported into simulation programs such as COMSOL [31]. For Iber et al., this revealed that Turing patterns were sensitive to the initial conditions of UB shapes [27–29]. Adivarahan et al. concluded that the Turing mechanism and the morphogen, such as glial cell-derived neurotrophic factor (GDNF), and diffusion-dependent geometry effect play an important role in the kidney branching morphogenesis [27].

As for the mechanism of the UB induction, Lawson and Flegg recently discovered it to be governed by Schnakenberg-like ligand-receptor Turing patterning [3]. The model explored the roles of Gremlin 1-like protein (GREM1) and bone morphogenetic protein (BMP) together with the domain properties of GDNF productions. The model considered the concentrations of BMP4, GREM1, BMP receptors, GDNF, and Ret proto-oncogene (RET). It devised one-by-one individual partial differential equations (PDE) to all of these concentrations. PDEs were dependent on time, place, production and degradation of these factors, and their binding to the respective receptors. The model non-dimensionalized PDEs, assumed zero initial concentrations, used mostly non-experimentally known values for the initial parameters, and numerically solved and simulated the PDEs with MATLAB. The model showed how these concentration levels change next Wolffian duct in such a way that UB is induced in specific position at further stage of WD especially affected by modulations in GDNF and RET levels.

In the same fashion, some concerns have been also given to renal interstitial fibrosis modelling [8, 9, 32]. These papers suggested the importance of a careful and comprehensive analysis of the underlying molecular interactions before proposing a Turing mechanism and recommend geometry and signaling to be reasons as such for the patterning. Conversely, the writers did not associate these reasons for UB development to the mechanochemical models. Indeed, the mechanochemical analysis of UB development has not been well studied, also since it is beyond the classical Turing explanation for the tissue curvature and the morphogen expression [26]. Accordingly, all these findings implicate and highlight the importance of a more comprehensive understanding of mechanochemical interactions and signaling in the kidney development [26]. This kind of an analysis would suggest that the branching and morphogen expressions levels are coupled in an inherently positive feedback loop instead of a long-range inhibitor of the previous model [26].

## 3 The Examples

### 3.1 An Introduction

Due to time scale of the differentiation of MM, the aging of NPs should be less important to the patterning propensities than the adhesion and extracellular elements [33, 34]. The hypothesis for the NP movement model (NPM) was that the simulated 2D process data would follow the cell migration data of the earlier ex vivo experiments [19, 23, 35, 36]. The aggregation of tubules to vesicle was named to a PTA to RV model (PTA-RV). The cell movements and patterning for two incipient nephrons were assessed, since the appearance of nephrons occurs at various developmental times below UB tips. As a result, the complex NP sorting process that involves diffusing extracellular chemoattractant and other cells; a multiscale modelling approach, such as probabilistic Cellular Potts Model (CPM); and modelling was needed. CPM simulations were performed in CompuCell3D program and analyzed with python programming. The realistic parameters for the program were found from the literature or by estimating with parameter scans [37, 38]. These computational analysis examples described important aspects of the early kidney development modelling [10, 39].

### 3.2 Cellular Potts Model

CPM represents all the cells as regions or domains within a regular lattice. The cells in this virtual Petri dish in the start of the in silico process simulation resemble boats in the battleship game. CPM designates an index for these cell domains, $\sigma_{\bar{I}}$ or $\sigma_{\bar{J}}$. This regularity and apparent uniformity of the model enables simulation of various structures, such as interactions between cells, adhesion $\bar{J}$, and diffusion in extracellular space, $D$ [40]. The model matches the forces to a generalized energy, $H$. It has been represented as a cost function consisting of sums of the interfacial energies, i.e., a cortical tension and adhesion, and energies due to the deviation of cell from a resting volume, $v$. CPMs treat cells phenomenologically reducing the millions of interactions of gene products to several behaviors. These behaviors include moving, dividing, dying, differentiating, changing shapes, exerting forces, secreting and absorbing chemicals and electrical charges, and changing their distribution of surface proteins [39]. Interactions and dynamics define how CPM objects behave both biologically and physically. The cell behaviors are embodied in the effective energy, which determines the cell shape, motility, adhesion, and response to extracellular signals.

The CPM simulation computations contain a stochastic component. This arrives from the dynamic Monte Carlo simulation algorithm. It tries to minimize the free energy between two adjacent time points, Monte Carlo steps, MCS. The algorithm randomly selects a cell lattice pixel and attempts to copy its index into a randomly chosen pixel on the border of another neighboring cell. It then checks if the site belongs to a different index, i.e., to different cell types. The site change happens automatically in the case of different cells if the net energy change is negative, $\Delta H \leq 0$. However, it violates the second law of thermodynamics for the confined systems and is in addition a very unlikely event in the real nature. Therefore, the cell transition likelihood, $P$, to neighboring site of a different cell must follow a different rule

$$P = \left\{ \begin{matrix} 1 & \text{if } \Delta H \leq 0, \\ e^{(-\Delta H/T)} & \text{if } \Delta H > 0 \end{matrix} \right\}, \tag{1}$$

where $T$ can be depicted as the magnitude of membrane fluctuations and in that fashion as a temperature, $T$. In effect, the probability (1) that the index copy change is accepted for the other case, $\Delta H > 0$, follows the Boltzmann distribution [14]. A density (pcs/m) of a floating secreted protein, e.g., Wnt9b, is represented by a function $u(t, x)$ at time $t$ and point $x$. The diffusion equation arrives by deriving the function $u$ back to its partial differential equation form

$$\frac{\partial}{\partial t} u = \mu \frac{\partial^2}{\partial x^2} u + f(u) \tag{2}$$

where the function $f(u)$, the reaction term, and the diffusion constant $\mu$ should be determined experimentally. The derivate part is called the diffusion term, for as well called the reaction-diffusion equation. These terms can be estimated henceforth with the literature [36, 39].

### 3.3 Nephron Progenitor Cell Movement Model (NPM)

The moving patterns of cells in the early phases of normal nephrogenesis consist of many synergistic biological mechanisms [8]. To add the complexity of the study, the proper mathematics describing large cell movement patterns have been scarcely characterized in the literature [41–43]. These instances mention, e.g., Gilbert-Johnson-Keerthi algorithm, nonlinear diffusion-reaction equations, and the Fürth equation as exemplary methods for describing the cell moving and collision experiments and simulations. For instance, the extracellular and intracellular forces can identically affect to the collective movement. Therefore, these forces may be shortly delineated in a theoretical way as tension fields; concentrically as internally affecting myosin contraction, actin polymerization, and volume-regulating processes; or externally as adhesion molecules [44]. If the cell-secreted morphogens from NPs, UBs, and the interstitial medium cells are to be counted, such as Wnt9b and Bmp7, the description of the mechanism gets even more complicated [7, 12, 45]. In these respects, the scope of possible kinetical formulations around nephron cell migration and signaling processes would be vast [46]. As an example, without transcription factors of Eya1, Cited1, Hox11, Osr1, Pax2, Sall1, Six1, Six2, and Wt1, there would not be normal kidney development. In particular, the elimination of genes other than Six2 and Sall1 causes the loss of progenitor cells and utter absence of nephrons. Furthermore, the survival of NPs is enabled by Bmp7 arriving from MM cells [7].

Consequently, the computational model of the movements of the NPs (NPM) toward PTA location in the UB bend was conceived to originate from just two main motifs: on the one hand, from the self-generated chemotaxis and on the other hand from the membrane differences of these cells compared to the other surrounding cells [8, 33, 45, 47]. The practical construction of NPM was derived from a cell sorting example of the CompuCell3D program in 2D [8, 14, 36]. Besides NPs, NPM consisted of interstitial medium, uninduced MM cells, and two UBs. It kept the UB cells static while the evaluation time of the process was short compared to the overall kidney development, and UB is big compared to NPs (Fig. 1). The speeds of NPM indicated a relationship to the analogous MM cell swarming around UB tips [8, 14].

### 3.4 Pretubular Aggregate to Renal Vesicle Model (PTA-RV)

For depicting the transition of PTA to RV, the computational construction of the model was also done with CompuCell3D [39]. However, the assumptions and model parameters were quite different [48], namely, the secretion of β-catenin arriving from the NPs alongside with small intrinsic cellular division and elongation and the differential secretion of Wnt9b and Fgf8 coming from UB corner. The amount of initial NPs and parameters were derived from the literature; in 3D there are 100 which is about 20 in 2D if uniformly distributed [39]. The initial construction has been depicted in Fig. 2. The upper right corner of the simulation lattice

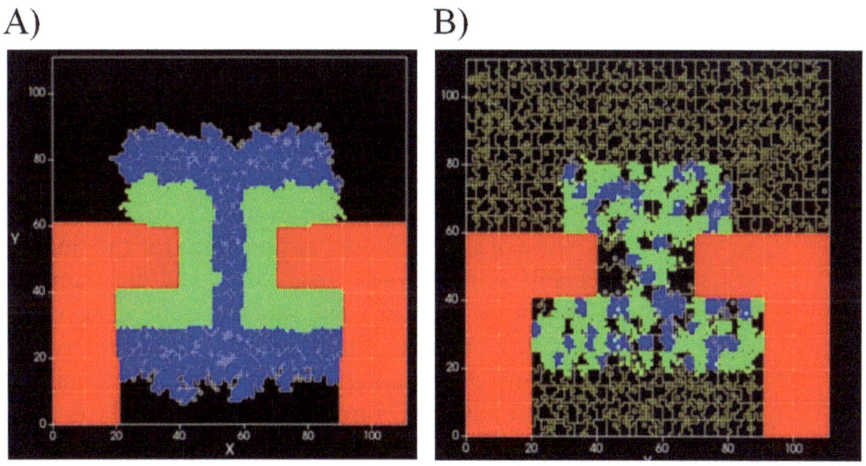


**Fig. 1** The initial constructions, **a** and **b**, of the cellular domains at CompuCell3D program for the nephron progenitor movement model. The nephron progenitors are green, ureteric buds are red, metanephric mesenchymal cells are blue, and the medium is black. The virtual lattice dimensions of the constructions are 111×111 in the x–y coordinates. The individual cell positions of the hypothesized start formations of the biological process were defined numerically with Excel calculations. In the **b** case (contrary to **a**), the positions of non-UB cells were pseudo-randomized in realistic regions to find such cells. Equally in the **b** case, the outlines of media cells were visually depicted as to show that they are really there

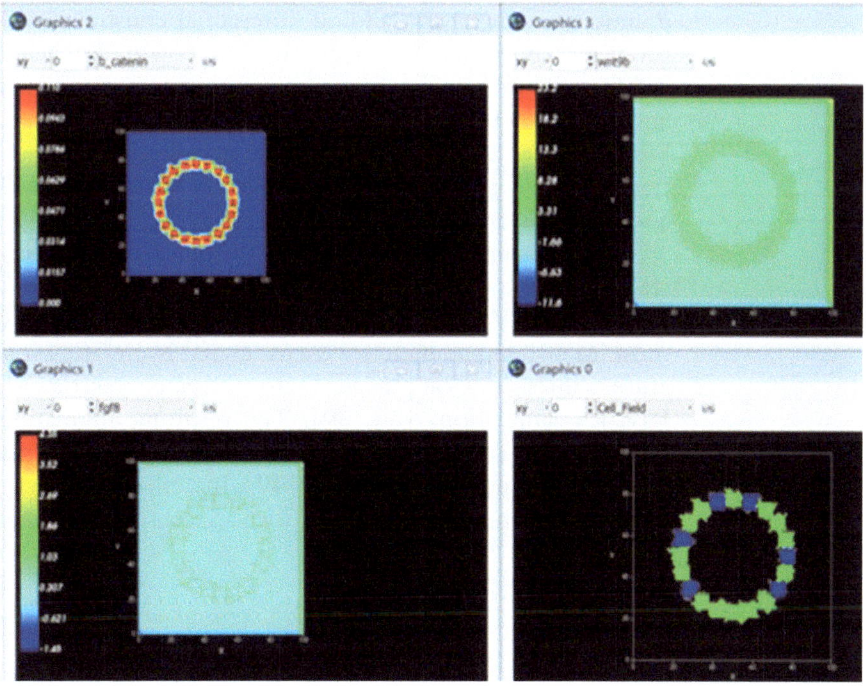


**Fig. 2** The initial construction of PTA-RV at CompuCell3D. Upper left, the initial amount of β-catenin from the nephron cells; upper right, the amount of Wnt9b from ureteric bud; lower left, the amount of Fgf8; and lower left, the initial cellular construction of the PTA-RV model. Nephron progenitor cells are green, and six MM cells, blue, were allowed to participate in the model

represents UB. The shape of the construction at the end of the simulation resembled RV.

## 4 Modelling NPM and PTA-RV with CompuCell3D and Python Analysis Codes

Constructing a cellular model in CompuCell3D requires a number of steps. First of all, the model assumptions for the simulation have to be clearly written down. In essence, the program input parameters have to be evaluated and specified so that the output image of the model simulation correspond the outcome of the real process. The assumptions are inserted to the model building wizard system of CompuCell3D program that templated the model files of the program. These three files, one xml and two python files, exacted specific information about the model. The Xml file requested the user-specified names of the cells, e.g., Medium, NPs, MMs, and UB.

Depending on the process or outcome, the program inquires cellular initial and end volumes and lambda constants, contact energies, secretion factors, and Wnt9b diffusion constants to the partial differential equation solver, DiffusionSolverFE. And finally, the initial construction, i.e., piff file location, was requested. Python files were in the light of revised, since the power of programming comes from adding differential constraints to the cell properties after each MC steps, such as growth, and cell tracking. The general python file, for example, Name.py, is for maintaining and compiling the efficient running of the second one, e.g., Name-Steppables.py. The later file inquired the nephron progenitor cell tracking plot parameters, plot's file location, and simulation start and running variables, the cell volumes.

It was assumed that there is no division, and cells do not ask for steering of the chemotaxis flow toward any particular cell type in NPM. PTA-RV had a small cell growth from 80 to 200 cells. The most important parameters were estimated by iterating initial literature values in CompuCell3D to recreate realistic NPM and PTA-RV simulations [36, 39]. The resulting simulation file of compucell3D can be analyzed together with the experimental file in Spyder program using python [49]. In general, it is a good habit to keep two separate files, one for the variables and the other for the analysis.

### References

1. Cartwright N (1983) How the laws of physics lie. Oxford University Press, Oxford
2. Berg VD (2011) Mathematical models of biological systems, Oxford University Press, Oxford
3. Lawson B, Flegg M (2016) A mathematical model for the induction of the mammalian ureteric bud. J Theor Biol 394:43–56
4. Short KM, Combes AN, Lefevre J et al (2014) Global quantification of tissue dynamics in the

developing mouse kidney. Dev Cell 29:188–202
5. Junttila S, Saarela U, Halt K et al (2014) Functional genetic targeting of embryonic kidney progenitor cells ex vivo. J Am Soc Nephrol 26:1126–1137
6. Takasato M, Little MH (2015) The origin of the mammalian kidney: implications for recreating the kidney in vitro. Development 142:1937–1947
7. Krause M, Rak–Raszewska A, Pietilä I et al (2015) Signaling during kidney development. Cells 4:112–132
8. Zubkov VS, Combes AN, Short KM et al (2015) A spatially-averaged mathematical model of kidney branching morphogenesis. J Theor Biol 379:24–37
9. McMahon AP (2016) Development of the mammalian kidney. Curr Top Dev Biol 117:31–64
10. Little M et al (2015) Kidney development, disease, repair and regeneration, 1st edn. Academic Press, London
11. Karner C et al (2011) Canonical Wnt9b signaling balances progenitor cell expansion and differentiation during kidney development. Development 138:1247–1257
12. Graner F, Glazier JA (1992) Simulation of biological cell sorting using a two-dimensional extended Potts model. Phys Rev Lett 69:2013–2017
13. Hirashima T, Elisabeth R, Roeland M (2017) Cellular Potts modeling of complex multicellular behaviors in tissue morphogenesis. Dev Growth Amp Differ 59:329–339
14. Turing AM (1952) The chemical basis of morphogenesis. Philos Trans Roy Soc London 237 (641):37–72
15. Gierer A, Meinhardt H (1972) A theory of biological pattern formation. Kybernetic 12:10–39
16. Schnakenberg J (1979) Simple chemical reaction systems with limit cycle behaviour. J Theor Biol 81:389–400
17. Combes A, Lefevre J, Wilson S, Hamilton N et al (2016) Cap mesenchyme cell swarming during kidney development is influenced by attraction, repulsion, and adhesion to the ureteric tip. Dev Biol 418:297–306
18. Meinhardt H, Gierer A (1980) Generation and regeneration of sequence of structures during morphogenesis. J Theor Biol 85:429–450
19. Kohonen T (1982) Self-organized formation of topologically correct feature maps. Biol Cybern 43:59–69
20. Murray JD, Oster GF, Harris AG et al (1983) A mechanical model for mesenchymal morphogenesis. J Math Biol 17:125–129
21. Murray JD, Maini PK (1988) Mechanochemical models for generating biological pattern and form in development. Phys Rep 2:171
22. Koch AJ, Meinhardt H (1994) Biological pattern formation: from basic mechanisms to complex structures. Rev Mod Phys 4:66
23. Plaza RG, Sánchez–Garduño F, Padilla P et al (2004) The effect of growth and curvature on pattern formation. J Dyn Diff Equat 4:16
24. Mercker M, Hartmann D, Marciniak–Czochra A (2013) A mechanochemical model for embryonic pattern formation: coupling tissue mechanics and morphogen expression. PLoS One 8:12
25. Mercker M, Köthe A, Marciniak-Czochra A (2015) Mechanochemical symmetry breaking in hydra aggregates. Biophys J 108:2396–2407
26. Mercker M, Brinkmann F, Marciniak–Czochra A et al (2016) Beyond Turing: mechanochemical pattern formation in biological tissues. Biol Direct 22:11
27. Adivarahan S, Menshykau D, Michos O et al (2013) Dynamic image-based modelling of kidney branching morphogenesis. Swiss Inst Bioinform 1:30100305
28. Menshykau D, Dagmar D (2013) Kidney branching morphogenesis under the control of a ligand-receptor-based Turing mechanism. Phys Biol 10:046003
29. Menshykau D, Iber D (2013) The control of branching morphogenesis. Open Biol 3:130088
30. Inc, Mathworks (2018) Matlab program. https://www.mathworks.com/
31. Inc, Comsol. (2018) Comsol program. https://www.comsol.com/
32. Hao W, Rovin BH, Friedman A (2014) Mathematical model of renal interstitial fibrosis. PNAS 39:111
33. Lefevre JG et al (2017) Self-organisation after embryonic kidney dissociation is driven via selective adhesion of ureteric epithelial cells. Development 144(6):1087–1096
34. Saxén L (1987) Organogenesis of the kidney. Cambridge University Press, Cambridge
35. Carter (2009) Visual group theory. Mathematical Association of America, Washington, DC
36. Andasari V, Roper RT, Swat MH et al (2012) Integrating intracellular dynamics using CompuCell3D and Bionetsolver: applications to multiscale modelling of cancer cell growth and invasion. PLoS One 7(3):e33726

37. Little MH (2015) The life cycle of the nephron progenitor. Dev Cell 35:5–6
38. Belmonte JM, Clendenon SG, Oliveira GM et al (2016) Virtual-tissue computer simulations define the roles of cell adhesion and proliferation in the onset of kidney cystic disease. Mol Biol Cell 27:3673–3685
39. Swat MH et al (2009) Multicell simulations of development and disease using the CompuCell3D simulation environment, vol 500, pp 361–428
40. Lorenzi T, Lorz A, Perthame B (2017) On interfaces between cell populations with different mobilities. Kine Relat Models 10:299–311
41. Tweedy L, Knecht D, Mackay G (2016) Self-generated chemoattractant gradients: attractant depletion extends the range and robustness of chemotaxis. PLoS Biol 14:e1002404
42. Magno R, Grieneisen VA, Athanasius FMM (2015) The biophysical nature of cells: potential cell behaviours revealed by analytical and computational studies of cell surface mechanics. BMC Biophys 8:2046–1682
43. Kofahl B, Wolf J (2010) Mathematical modelling of Wnt/ β–catenin signalling. Biochem Soc Trans 38:1281
44. Blessing OE et al (2015) A mathematical model of quorum sensing induced biofilm detachment. PLoS One 10:e0132385
45. David E et al (2016) Physical forces shape group identity of swimming pseudomonas putida cells. Front Microbiol 7:1437
46. Albert PJ, Schwarz US (2016) Dynamics of cell ensembles on adhesive micropatterns: bridging the gap between single cell spreading and collective cell migration. PLoS Comput Biol 12: e1004863
47. Murphy PM (2010) Double duty for CCL21 in dendritic cell trafficking. Immunity 32:590–592
48. Marciano DK (2016) A holey pursuit: lumen formation in the developing kidney. Pediatr Nephrol 32:7–20
49. Spyder (2017) Spyder program. https://pythonhosted.org/spyder/installation.html

MIX
Papier aus verantwortungsvollen Quellen
Paper from responsible sources
FSC® C105338

If you have any concerns about our products,
you can contact us on
**ProductSafety@springernature.com**

In case Publisher is established outside the EU,
the EU authorized representative is:
**Springer Nature Customer Service Center GmbH**
**Europaplatz 3, 69115 Heidelberg, Germany**

Printed by Libri Plureos GmbH
in Hamburg, Germany

# Index



Seppo Vainio (ed.), *Kidney Organogenesis: Methods and Protocols*, Methods in Molecular Biology, vol. 1926, https://doi.org/10.1007/978-1-4939-9021-4, 